Fate of Chemicals
in the Environment

Fate of Chemicals in the Environment

Compartmental and Multimedia Models for Predictions

Robert L. Swann, EDITOR
Dow Chemical Company

Alan Eschenroeder, EDITOR
Arthur D. Little, Inc.

Based on a symposium

sponsored by the ACS Division

of Pesticide Chemistry

at the 184th Meeting of the

American Chemical Society,

Kansas City, Missouri,

September 12–17, 1982

A C S   S Y M P O S I U M   S E R I E S **225**

AMERICAN CHEMICAL SOCIETY
WASHINGTON, D.C. 1983

Library of Congress Cataloging in Publication Data

Fate of chemicals in the environment.
  (ACS symposium series, ISSN 0097–6156; 225)

  Bibliography: p.
  Includes index.

  1. Chemicals—Environmental aspects—Mathematical models—Congresses. 2. Environmental chemistry—Mathematical models—Congresses. I. Swann,
Robert L., 1944–   . II. Eschenroeder, Alan, 1933–
      . III. American Chemical Society. Division of Pes-
ticide Chemistry. IV. Series.

TD196.C45F37 1983   628.5   83–12209
ISBN 0–8412–0792–5

ACS Symposium Series

M. Joan Comstock, *Series Editor*

FOREWORD

The ACS Symposium Series was founded in 1974 to provide a medium for publishing symposia quickly in book form. The format of the Series parallels that of the continuing Advances in Chemistry Series except that in order to save time the papers are not typeset but are reproduced as they are submitted by the authors in camera-ready form. Papers are reviewed under the supervision of the Editors with the assistance of the Series Advisory Board and are selected to maintain the integrity of the symposia; however, verbatim reproductions of previously published papers are not accepted. Both reviews and reports of research are acceptable since symposia may embrace both types of presentation.

CONTENTS

HUMAN RISK ASSESSMENT

PREFACE

PUBLIC CONCERN OVER THE EFFECTS of chemical release into the environment through human activity has grown steadily since the appearance of Rachel Carson's "Silent Spring." This concern focuses not only on the potential threats to human health, but also on indirect harm arising through disruptions of natural ecosystems. Even when economic hardship challenged our desire to preserve the quality of air, water, and foodstuffs, persistent demands to satisfy that desire were expressed in the form of public laws, regulations, and enforcement activities covering a broad array of topics. Industry has taken the lead in several areas in anticipation of regulations and in awareness of the higher social cost compared to the prevention cost.

The need to balance costs against benefits both in the public and private sectors resulted in a search for methods of predicting the fate and effects of chemicals in the environment. Actual field testing of all cases of interest is both too costly and too dangerous to perform. Mathematical models, therefore, have been developed to provide descriptive tools and predictive approaches to this problem. At the symposium on which this book is based, a collection of user-oriented information was presented and covered the following aspects of environmental fate modeling:

1. The needs motivating development of each class of model.
2. The theoretical background underlying the structure of a mathematical scheme for simulating the phenomenology of processes in air, water, soil, biota, or a combination of media.
3. A generic overview of the main building blocks of models and the framework of logic connecting the components.
4. An operational description of model applications and user influence on model choice.
5. Case studies of specific model applications.

The symposium blended tutorial review papers with descriptions of field, laboratory, industrial, and regulatory problems that have been approached using chemical fate simulations. Authors presented current practices and practical questions such as material balance analysis, atmospheric processes influencing human exposure, aquatic system pathway analysis, movement in soil/groundwater media, and uptake or degradation in biota.

The editors believe that this book fills a void in the coverage and should stimulate further research in this vital area of the environmental sciences.

ROBERT L. SWANN
Dow Chemical
Midland, Michigan

ALAN ESCHENROEDER
Arthur D. Little
Cambridge, Massachusetts

May 10, 1983

CHEMICAL RELEASE

Release of Chemicals into the Environment

STEPHEN L. BROWN and DAVID C. BOMBERGER

SRI International, Menlo Park, CA 94025

This paper is a review of methods for estimating
releases of chemicals into the environment in the
course of extraction of raw materials,
manufacturing, use, storage, transportation, and
disposal, as well as by accidents or natural
processes. It discusses source types, forms of
substances released (solids, liquids, and gases),
receiving media (air, water, soil), time pattern
of release (continuous versus intermittent,
cyclic versus random), and geographic patterns
of release (point, line, area, and volume sources).
The paper reviews several ad-hoc approaches to
estimating releases and illustrates their use
with a case study of benzene. The authors identify
key opportunities for further research.

This symposium concerns models for predicting the fate of
chemicals in the environment. Strictly speaking, the topic of
this paper does not fall into the usual definition of fate
models. However, every fate model has at least one source term.
Although the source term for one fate model may be the output of
another fate model (as when air transport models provide the
deposition rates that are the inputs to an aquatic fate model),
the chain always has to be traced to the original sources,
whether they are natural or associated with human activities.
In this paper, we characterize the various sources for chemicals
in the environment and discuss methods for describing the
releases from them in terms sufficiently quantitative for use by
fate models.

We first describe human activities that can cause releases
of chemicals; these are usually of greatest concern to fate
models, because they suggest where interventions can be made and
environmental concentrations can be reduced. We then classify
releases by their form, medium of entry, and spatial and temporal
patterns. After briefly noting the most usual quantitative

0097–6156/83/0225–0003$06.00/0

expressions of release, we discuss several approaches to
estimating these quantities. Finally, we describe an ad-hoc
approach for an example chemical and note some areas for fruitful
future research.

<u>Human Activities That Cause Releases</u>

Chemicals are distributed in the environment by a wide variety of
natural processes, including physical (e.g., weathering),
chemical (e.g., photochemical), and biological (e.g.,
respiration) processes. Although many of these processes are
best thought of as closed cycles, not entailing a true "source,"
many can be thought of as source to sink processes, such as the
release of carbon dioxide by volcanic action and its sink in
oceanic carbonates. These natural processes form important
background source terms for chemicals, but they are usually not
of primary interest because they are often not controllable.
Human activities that release chemicals, however, are of primary
interest, because some chemicals are released, in fact created,
solely by human activities and would not otherwise be found in
the environment.
    Without trying to make an exhaustive list of all the types
of human activities that cause releases, we can list many
different activities that are distinct and significant. Figure 1
shows a selection of such activities, indicating how they are
connected through the life cycle of a chemical and the media to
which they most commonly cause releases.
    The life cycle of some chemicals begins with extraction of
raw materials. Activities such as coal and mineral mining, oil
production, and forestry can either release chemicals directly or
open the land for releases by natural processes that otherwise
would be slower.
    Sometimes chemicals are prepared for distribution without
chemical reactions, as when limestone is mined and refined before
use. In other cases, the raw materials are converted to other
chemicals in a manufacturing process. In both cases, wastes are
discharged to air, water, and (if large quantities of solid or
semi-solid wastes are involved) to land.
    Both before and after processing, chemicals are stored and
transported, often many times and through many stages of
processing and manufacturing. Both storage and transportation
can entail "normal" low levels of release and occasional high
levels of release from accidents. (Manufacturing upsets also can
cause major accidental releases, such as the release of dioxins
from a trichlorophenol reactor at Seveso, Italy in 1976.)
    The major release mechanism for many chemicals, however, is
associated with use of the chemical or of chemical-containing
products. Primary uses include combustion of fuels, industrial
uses, commercial uses, household and other consumer uses,
deliberate applications in the environment (for example,

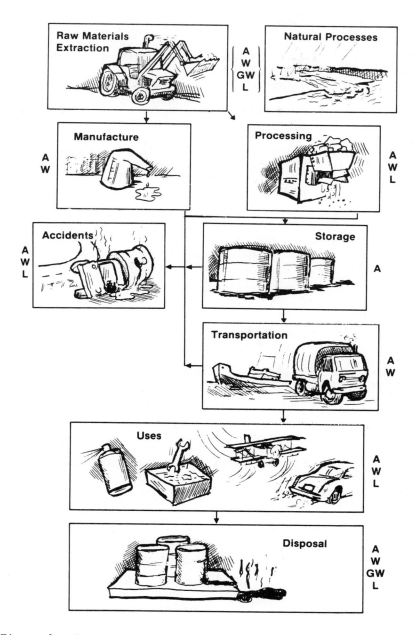

Figure 1. Human activities leading to release of chemicals into the environment. Key: A, air; W, water; GW, groundwater; L, land.

pesticides), and many others. In some cases (spray can
propellants, for example), virtually all of the chemical used is
released to the environment. In other uses (such as in vinyl
asbestos floor tiles), most of the chemical constituents are
essentially isolated from the environment for long periods of
time. Such isolated reservoirs are also "sinks" for the
chemical.

Even with such uses as floor tiles, however, the time
eventually comes for disposal, and materials find their way into
public sewage systems, dumps, and other less-formal disposal
facilities. Secondary processes release chemicals from these
facilities into all environmental media.

Form of Substances Released

Chemicals may be released in solid, liquid, or gaseous forms and
release may be to air, surface water, groundwater, and land.
(We include rivers, lakes, and the oceans as surface waters; both
interstitial water in soils and deeper aquifers as groundwater;
and application to soil as well as shallow or deeper burial as
land releases. The distinction between a land release and a
groundwater release is largely arbitrary.) The forms in which
the chemicals appear in those media are varied, as shown in
Table I.

Table I

FORMS OF SUBSTANCES IN RECEIVING MEDIA

Media	Solids	Liquids	Gases	Combinations
Air	Particulate	Vapor Particulate	Gas	Adsorbed gas or liquid
Water	Suspended Dissolved	Dissolved	Dissolved	Cosolution
Groundwater	Dissolved	Dissolved	Dissolved	
Land	Particulate Bulk Contained	Contained Absorbed Adsorbed	Contained	

In air, solids appear as particulates, liquids as either
particulates or vapors, and gases of course in gas form.
Combinations are also possible, as when gases are adsorbed on
particulates. Some solids would also have substantial vapor
pressures, and so on, but we have tried to simplify the exhibit.
In water, a dissolved state is typical for substances that are
normally solid, liquid, or gaseous, but solids can also simply be

suspended. Groundwater is somewhat less likely to contain
dissolved gases, but that too is possible. Land receives solids
as scattered particulates or in bulk as well as in containers.
Liquids may fill interstitial voids of otherwise dry soils, be
adsorbed to soil particles, or remain contained. Gases can also
be contained or adsorbed. Whether "release" occurs when
containers are placed in landfills or not until after they are
breached is, again, largely a matter of definition.

Source Characteristics

Environmental fate models require information on the distribution
of releases over time and space. Basically, sources can be
described in terms of their dimensionality and releases in terms
of their temporal distribution.

Dimensionality is best illustrated by defining sources of
air pollution (Figure 2). Point sources, such as the mouths of
smokestacks, release pollutants at (almost) a single point in
space, which can be described by its geographic coordinates and
height above the surface (or above sea level). Line sources are
unidimensional, although they do not have to be straight lines;
for example, on a roadway, cars form moving point sources that,
in aggregate, look much like a nearly uniform line source. As
represented in Figure 2 by only one house, a group of residences
burning wood for heat can often be better treated as a two-
dimensional area source than as a large set of point sources; the
distribution of gas stations in an urban area is also probably
sufficiently well simulated by an area source.

Our concept of a volume source (see Figure 2) is
intentionally vague, because few good examples exist. However,
photochemical smog is produced over a volume of air; is this just
part of a fate model or should it be considered a source?

Line, area, and volume sources are also described by their
geographic distribution, shape, and orientation. For surface
water, an outfall is a point source, whereas runoff to a river is
a line source and deposition from the air is an area source.
Similar ideas can be applied to the groundwater and land media.

There are also several possibilities for the temporal
distribution of releases. Although some releases, such as those
stemming from accidents, are best described as instantaneous
release of a total amount of material (kg per event), most
releases are described as rates: kg/sec (point source), kg/sec-m
(line source), $kg/sec-m^2$ (area source). (Note here that a little
dimensional analysis will often indicate whether a factor or
constant in a fate model has been inadvertently omitted.) The
patterns of rates over time can be quite diverse (see Figure 3).
Many releases are more or less continuous and more or less
uniform, such as stack emissions from a base-load power plant.
Others are intermittent but fairly regular, or at least
predictable, as when a coke oven is opened or a chemical vat

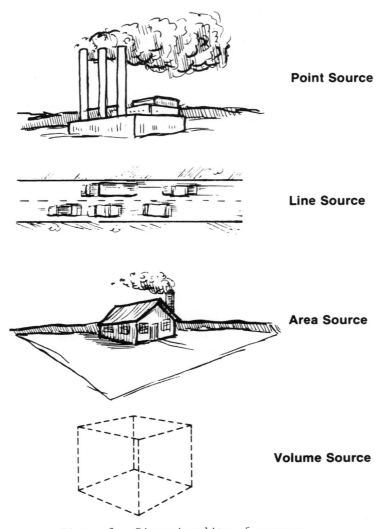

Point Source

Line Source

Area Source

Volume Source

Figure 2.   Dimensionality of sources.

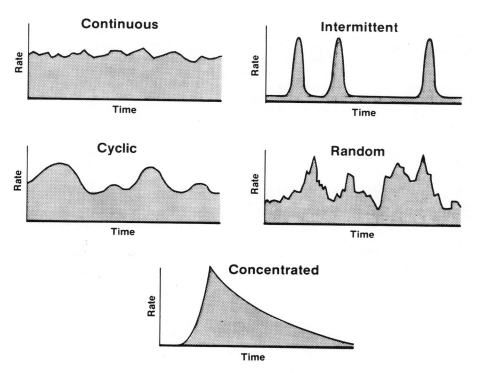

Figure 3.   Time patterns of release.

purged. Some are continuous but cyclic, such as automobile
emissions over a day, and some are more or less random, either
continuous or intermittent, as might occur when rain falls into
a waste treatment pond and causes it to overflow into surface
waters. Accidental releases, of course, are concentrated into
individual events that, nonetheless, may cause releases
persisting over a period of time.

These temporal patterns are characterized by a variety of
quantitative measures of the rates of release. Any pattern, of
course, can be described in detail as a function of time [$r(t)$ =
1 kg/sec, t = 8 am to 5 pm; $r(t)$ = 0 otherwise]. However, it is
often sufficient to characterize some typical rate or one of
special interest. In air pollution, annual average emission
rates are often sufficient if the goal is to predict annual
average concentrations. But if the highest 24-hour, 8-hour,
3-hour, or 1-hour averages for concentrations are desired,
similarly time-segregated emission rates may be needed (see
Figure 4). For some of the intermittent or accidental releases,
it may be sufficient or even desirable to give integrated
releases, wherein the release rates are integrated over some
time of interest. Such cases also may be approximated by
equivalent constant release rates over the same time period.

Approaches to Estimating Releases

The preceding descriptions make it very easy to characterize how
release information is desired; unfortunately, however, it is not
so easy to estimate such quantities from readily available
information. Some of the major types of estimating techniques
are illustrated in Figure 5.

All release information can be tracked back to measurement,
and direct measurement is frequently the preferred way of
estimating emissions. Stack gas sampling is a case in point: we
measure concentrations in stack gas, measure gas flow rates, and
compute emission rates with essentially no error other than that
caused by inaccurate instruments or insufficient samples to
characterize a full annual sample. Other examples are automotive
exhaust sampling, discharge pipe sampling (aqueous effluents),
and manifests for land disposal by weight and volume. Measured
disappearance rates for storage or transportation can be inferred
to be release rates. Application rates for pesticides and
fertilizers are sometimes adequate surrogates for kg/m^2 release
rates to soils.

A second major estimating technique is the materials balance
approach--the original focus of this paper. A chemical
engineering standard, the materials balance can reduce to the
simple mass balance, as when the measured mass of a chemical in
products leaving the plant is subtracted from the raw material
entering the plant to yield the loss. This loss is then
partitioned among releases to various media or other sinks. If

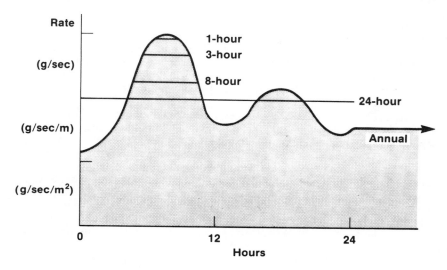

Figure 4.   Characteristics of release.

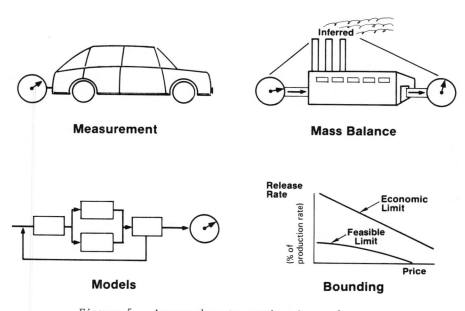

Figure 5.   Approaches to estimating releases.

chemical transformations are entailed, the technique becomes
known as a materials balance. More complicated balances can
start with the extracted or manufactured volume (kg/yr) of a
chemical and trace it to all its intended uses, including final
disposal. Proper accounting of all flows can yield important
information about the rates of release in various branches of
the distribution tree; however, relatively small uncertainties in
the product flows can cause huge relative uncertainties in the
release flows.

Mathematical models are also used for estimating releases,
but these are usually relatively simple. For example, if it is
known that X kg of a chlorofluorocarbon is manufactured annually,
and Y percent enters spray cans, and Z percent of a spray can is
usually left unexhausted, then $XY(100-Z)/10^4$ kg of that CFC are
released to the atmosphere per year. The average discharge rate
(kg/sec) nationwide then can be computed easily. (For simplicity
in this example, we ignore the contributions from leaking
discarded cans and changes in production and use levels.)

Other models might start with concentrations of heavy metals
in coal, amounts of coal used per kWh electricity produced, kW
capacity and load factor, scrubbing efficiencies, and so on to
produce an estimate of stack emission rates for a coal-fired
electric power plant. In these cases, the model starts with
measured quantities (production level, coal concentration), and
predicts the release rates. In other cases, an environmental
fate model is applied in a well-characterized situation and its
outputs are compared with measurements to calibrate the source
term. For example, a groundwater model can be developed with
semiempirical, adjustable constants, one of which is the source
strength. Sufficient comparisons of predicted groundwater
concentrations with concentration measurements from test wells
can achieve a good estimate of the source strength, which can
then be used to estimate concentrations at other places and times.

Most other techniques for estimating release rates are
ad-hoc, in the sense that one uses the most obvious suppositions,
calculations, and so on for a given situation. Some of these
techniques set bounds on release rates. For example, the
percentage of a manufactured product that will be tolerated as
waste depends on its price, because the profit margin can be
markedly degraded if too much is lost. At the same time, however,
high-profit materials can supply the resources to install more
effective controls. The results are rule-of-thumb bounds on the
fractions of production likely to be lost (see lower right
diagram of Figure 5). Another useful rule is that, given equal
prices, a smaller fraction will be lost from high-volume
processes than from low-volume ones. Continuous processes
inherently have smaller losses than batch processes, but there is
probably no such thing as a "completely closed process." From
such arguments, release rates would very rarely exceed 10% of
production for economic reasons, and very rarely would they be

lower than 0.01% of production on feasibility grounds. Typical
values might be in the range of 0.1% to 1%.

Example of an Ad-Hoc Approach: Benzene

Although benzene has recently come under increasing control
because of its alleged role in leukemia and other neoplastic
diseases, in past years it has been widely dispersed in
commercial uses and has entered the environment through many
routes. In 1977, SRI attempted to characterize "Human Exposures
to Atmospheric Benzene" (1). Most of the following examples come
from that report, even though several later studies have updated
and refined that work, and recent events have changed exposure
patterns.
     First, points of release of benzene were identified:
petroleum refining and coke oven operations (production and
extraction releases), use as a chemical intermediate
(transportation, storage, use, and waste releases), use in
gasoline (use-related release), and use in finished products
(use-related release). Benzene also can be a contaminant of most
of the derivatives made from it and its use as a solvent was
substantial before health concerns arose. The complexity of the
chemical systems dependent on benzene is shown in Figure 6. A
list of potential releasing products appears in Table II.

Table II

MANUFACTURED PRODUCTS USING BENZENE AS A SOLVENT (1)

Rubber tires
Miscellaneous rubber products
Adhesives
Gravure printing inks
Trade and industrial paints
Paint removers
Coated fabrics
Synthetic rubber
Leather and leather products
Floor coverings

     Next, various quantitative techniques were used to estimate
releases by type of use. For use of benzene as an intermediate,
we relied on the "emission factor" technique, which estimates the
ratio of benzene release to total derivative production and then
applies this ratio to the production rate at specific locations.
Emissions factors were estimated from crude engineering
assessments of the chemical processes entailed (such as open
versus closed systems, continuous versus batch, and so on).
These crude estimates could be checked by comparing estimates of
ambient concentrations based on them against actual measurements

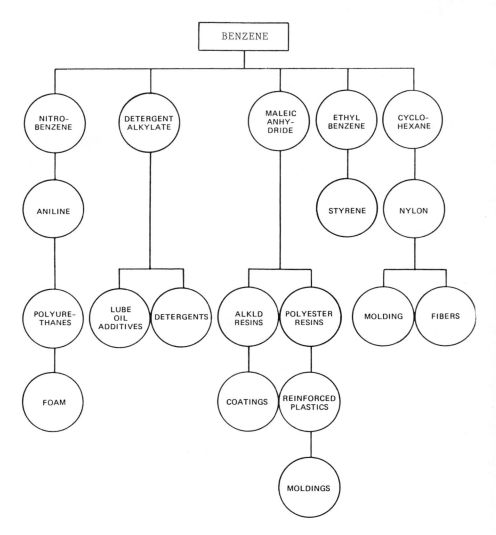

Figure 6a.   Benzene derivatives and their uses. (Reproduced with permission from Ref. 2.)

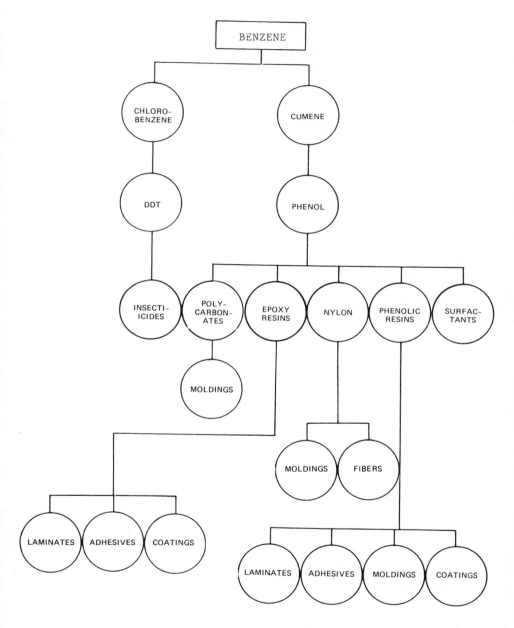

Figure 6b. Benzene derivatives and their uses. (Reproduced with permission from Ref. 2.)

or by mass balance on measured input and output. Even so, such
estimates were probably accurate to no better than a factor of
ten (Table III is a sample list). Emissions were then calculated
for about 80 plants, located in over 20 states, that consumed
about 7 billion pounds of benzene.

Table III

EMISSION FACTORS AND CHARACTERIZATIONS FOR
MANUFACTURING PLANTS THAT USE BENZENE

Chemical	Emission Factor (10^{-3} kg of benzene per kg of product)	Emission Characterization
SRI estimates (1):		
Aniline	23.60	Fugitive
Cyclohexane	0.25	Fugitive
Detergent alkylate (linear and branched)	2.20	Fugitive
PEDCo estimates (3):		
Cumene	0.25	Fugitive
Dichlorobenzene (p- and o-)	8.60	Chlorinator, by-product recovery systems
Ethylbenzene	0.62	Scrubber-vent
Maleic anhydride	96.70	Product recovery scrubber
Monochlorobenzene	3.50	Unknown
Nitrobenzene	7.00	Point absorber
Phenol	1.00	Unknown
Styrene	1.50	Collection vent, emergency vent

   Estimates also were made for 65 coke plants in 12 states.
Coke ovens produce benzene as a by-product, but not all of it can
be recovered. It has been estimated that benzene contributes
about two-thirds of one percent of the coal gas generated.
Potential points of emissions from one type of coke battery are
illustrated in Figure 7. Emissions from coke ovens were derived
from estimated emission factors (based on coke oven product
assays and benzene yields) and coal charging rates.
   Because the content in gasoline at that time accounted for
a large fraction of total benzene production, all parts of the
gasoline marketing chain (Figure 8) were considered to be

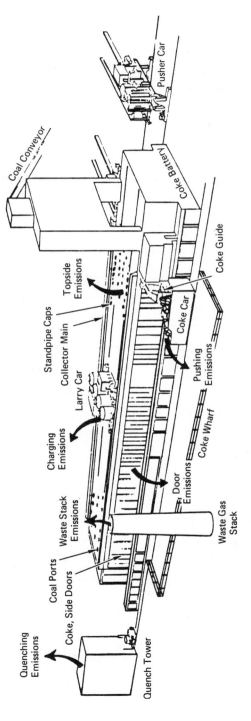

Figure 7. Schematic diagram of by-product coke oven showing possible atmospheric emission sources for Benzene. (Reproduced with permission from Ref. 3.)

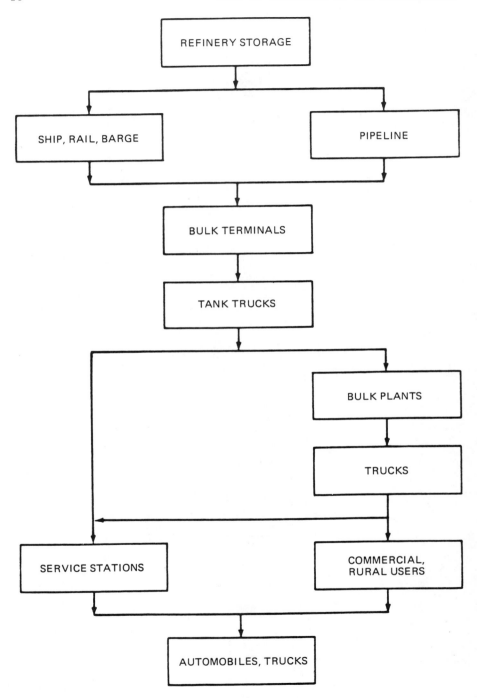

Figure 8.   The gasoline marketing distribution system in the
United States. (Reproduced with permission from Ref. 3.)

potential release points. Benzene can be emitted in refining, storage, dispensing (service stations), or use (automotive emissions). Refining releases were treated much as were those from use as an "intermediate," except that the emission factors were scaled to barrels of oil processed by using the estimated concentrations of benzene in total hydrocarbon emissions and emission factors per barrel of oil for hydrocarbons. Storage emissions are also based on emission factors. Both of the preceding source types were treated as point or small area sources. Service stations and automobiles, however, were treated as large area sources. Automotive tailpipe emissions were based on emission factors per mile driven and gas tank emissions were based on emissions per trip and trips per vehicle day. For service stations, ambient concentrations were predicted by models and compared with measurements to calibrate the emission rates.

For solvent operations, benzene use had to be estimated by a series of tenuous assumptions about the amount of benzene in "other uses," the percent of that used for solvents, and the loss of benzene from those operations. As an upper limit, it might be assumed that all of the purchased benzene is eventually lost to the atmosphere. However, some measured concentrations suggest that perhaps only 10% is lost at the plant. The remainder might be incinerated after becoming unusable or sent elsewhere for disposal. A general rule for volatile solvents is that they eventually reach the environment unless they are destroyed deliberately or degrade naturally. The distribution of solvent emissions geographically is much more difficult to determine.

For completeness, we must mention that benzene also occurs naturally in foods such as fruits, fish, vegetables, nuts, meats, dairy products, eggs, and alcoholic beverages. Exposures are estimated by multiplying measured concentrations by usage of the food product.

In the SRI report (2) the release information on benzene was used with atmospheric dispersion models and data on geographic distribution of population to obtain aggregate exposure estimates (shown in Table IV).

Table IV COMPARISON OF BENZENE EXPOSURES AMONG SOURCES (1)

Source	Exposure (10^6 ppb-person-years)
Chemical manufacturing	15.9
Coke ovens	8.8
Gasoline service stations	
People using self-service	1.6
People living in the vicinity	90.0
Petroleum refineries	3.4
Solvent operations	0.1
Storage and distribution	Minimal
Urban exposures from automobile emissions	102.2

Although several figures in Table IV are significant, the estimates are probably accurate to the first digit at best. However, they do suggest that widespread but low-level exposures from automobiles and service stations provide the majority of benzene molecules that enter human bodies. Whether these are the most biologically significant emissions depends on the behavior of dose-response relationships at low dose levels.

Research Opportunities

Because the techniques of estimating releases are so diverse and underdeveloped, there are many opportunities for improvement. However, the opportunities are difficult to describe in specific terms. We therefore note only a few broad areas:

. Measurement--Much uncertainty about release rates could be reduced by markedly increasing the number and variety of measurements made. Releases during use, if not clearly 100% of use or nearly so, are especially needed. We can clearly use many more model-measurement comparisons to calibrate our source term assumptions, as well as the model parameters.

. Statistics--We need better access to the data that are available from measurements. For example, annual production volumes are sometimes equal, or nearly so, to annual release rates on a nationwide basis. But concern for proprietary information has curtailed access to such data--in our opinion, out of proportion to the harm that might come to industry from public knowledge. Actual in-plant emission and effluent rates are obviously much more sensitive, but better summarization of distributions of such releases could be made for scientific use. Surveys of degree of use to combine with measured releases in such uses are also needed.

. Materials balance--This technique, in principle, is developed to its fullest extent, but it is extraordinarily sensitive to uncertainties in the data it uses. Better characterization of all pathways and chemical reactions would help, as would more accurate measurements of flows through these paths.

. Modeling--Most releases have been worked out with only one type of model. Variant approaches should be tried and compared. Opportunities should be sought to enrich models without overcomplicating them.

. Ad-hoc approaches--Methods of estimating should be borrowed from other problems whenever applicable. For example, statistical techniques for quality control theory can probably be applied to chemicals by viewing discharges as "faulty" production.

Literature Cited

1. Mara, Susan J., and Lee, Shonh S. "Human Exposures to Atmospheric Benzene," SRI International, Menlo Park, CA, 1977.
2. Hedley, W. M. "Potential Pollutants from Petrochemical Processes," Monsanto Research Corporation, 1975.
3. PEDCo Environmental "Atmospheric Benzene Emissions," PEDCo Environmental, Inc., 1977.

RECEIVED April 15, 1983.

ENVIRONMENTAL MODELS

Fate of Chemicals in Aquatic Systems: Process Models and Computer Codes

LAWRENCE A. BURNS

U.S. Environmental Protection Agency, Environmental Research Laboratory, Athens, GA 30613

Aquatic fate models are designed to forecast the residual concentrations, dominant transformation pathways, distributions among subsystems, and characteristic time scales of xenobiotic chemicals. Most are constructed as systems of differential equations organized around mass balances. The resulting computer codes are used as aids in chemical use and disposal evaluations; their primary function is to reduce complex chemical and environmental data sets to useful forms. Relevant chemical phenomena include direct and indirect photochemical reactions, hydrolytic processes, biotransformations, ionic speciation, and sorption. These phenomena include both reversible and irreversible processes, with a mixture of time scales ranging from the virtually instantaneous to the imperceptible, depending on the structure and reactivity of the chemical involved. Aquatic transport processes include hydrodynamic transport of dissolved materials, entrained transport of chemicals sorbed with particulates, volatilization, and exchange across the benthic boundary layer. The models combine chemical partitioning and rate constants with environmental driving forces, yielding a set of differential equations that can be analyzed to reveal chemical behavior as a function of time, space, and extrinsic chemical loadings.

When emitted to the aquatic environment, pesticides may endanger native populations and downstream drinking water supplies. A realistic evaluation of such dangers requires a knowledge of the transport and transformation processes that govern the exposure of organisms to chemical residuals. Exposure, fate, and persistence can be estimated via appropriately designed computer programs, thus providing an objective, rational, and replicable framework for chemical use and disposal decisions (1).

Aquatic fate models must account for several kinds of phenomena, including transport and transfer processes that move a compound among ecosystem segments and compartments, and degradation processes that convert the compound to transformation products. This chapter provides a brief summary of the quantitative basis for describing these processes, a description of the logic and data structures used to assemble fate codes, and a brief catalog of some of the main publicly available, better documented computer programs for aquatic fate.

Transport and Transfer Processes

Ionization, sorption, volatilization, and entrainment with fluid and particle motions are important to the fate of synthetic chemicals. Transport and transfer processes encompass a wide variety of time scales. Ionizations are rapid and, thus, usually are treated as equilibria in fate models. In many cases, sorption also can be treated as an equilibrium, although somtimes a kinetic approach is warranted (2). Transport processes must be treated as time-dependent phenomena, except in simple screening models (3,4).

Hydrodynamic Transport. Hydrodynamic transport is often modeled as a combination of advection and dispersion. Advection, the flow of water through the system, is usually represented via the mean flow velocity of a river, or hydraulic discharge from a lake or pond. Dispersion, an analogy with molecular diffusion, is a statistical accounting of the effects of turbulence, storm surges, internal waves, and other phenomena not amenable to detailed, mechanistic descriptions. The study of hydrodynamics is best developed for rivers (5) and estuaries (6), although the hydromechanics of ponds and lakes has not been totally neglected (7). Hydrodynamic flows carry an entrained load of dissolved materials. The transport of dissolved synthetic chemicals can be readily derived by regarding the flow as a simple carrier. In many instances a model for synthetic chemicals can be "piggy-backed" on either a hydrodynamic transport model or on a transport description derived from observational data.

Ionization. Many organic chemicals contain functional groups that dissociate to yield charged species. The toxicity and chemical reactivity of the uncharged (neutral) molecule and its charged ions can be very different. Differences in reactivity of ionic species can be accommodated in fate models when rate constants are expressed in terms of the individual species.

Ionization equilibria of acids and bases can be computed fairly easily. Consider, for example, an acid HA (H the hydrogen atom and A the remainder of the molecule). The distribution of HA between its un-ionized [HA] and ionized [A$^-$] species is controlled by the pH of the solution. Letting "I" denote the "ionization fractions" (8):

$$Io = 1 / (1 + Ka/[H^+]) = [H^+]/([H^+] + Ka) \qquad (1)$$

where Io is the fraction present as HA and Ka is the "ionization" or "acidity" constant. Mass conservation requires that the remainder be equal to (1 - Io); this fraction can also be expressed as $Ka/([H^+]+Ka)$. This concept of "ionization fractions" (or more generally, equilibrium species distribution fractions) can be extended to include the simultaneous occurrence of multivalent anions and cations, sediment-sorbed and biosorbed equilibrium states, etc.

Sorption. Capture of neutral organics by non-living particulates depends on the organic carbon content of the solids (9). Equilibrium sorption of such "hydrophobic" compounds can be described by a carbon-normalized partition coefficient on both a whole-sediment basis and by particle size classes. The success of the whole-sediment approach derives from the fact that most natural sediment organic matter falls in the "silt" or "fine" particle size fractions. So long as dissolved concentrations do not exceed 0.01 mM, linear isotherms (partition coefficients) can be used. At higher concentrations, the sorptive capacity of the solid can be exceeded, and a nonlinear Freundlich or Langmuir isotherm must be invoked.

Sorption of ionizable compounds (organic acids and bases) has yet to be described as a function of independent properties of compounds and environments. Ionization itself is quite fast and can be described as a simple equilibrium. The sorption of the resulting ions involves competition with normal inorganic and organic cations and anions. Sorption of organic cations can be described fairly well via a competitive Langmuir isotherm, using cation exchange capacity (CEC) as a general capacity factor (10). Given partition coefficients for neutral and ionic species, computation of the (local) equilibrium distribution of a compound is straightforward. Nonlinear isotherms and detailed sorption kinetics (2) have yet to be incorporated in operational exposure models, however, although some programs include a first-order kinetic approximation.

Particle Transport. Because many organic chemicals bind with aquatic particulate matter, particle transport can determine the fate of compounds. Sediment transport has been of interest to the engineering profession for many years. Many discussions of the dynamics of fluvial sediment transport have appeared in the literature (11, 12). As with hydrodynamic transport, one strategy for environmental modeling is to "piggy-back" the transport of sorbed chemicals on a model of transport of the sediment phase.

Interactions with Benthic Subsystems. In many lakes and coastal seas, net sediment deposition can be very slow. In these cases, capture of synthetic compounds may be driven primarily by the activities of the biota, including physical disturbance by demersal fishes, irrigation of the sediments by benthic macrofauna (13), and sorption to surface layers followed by subduction and mixing of contaminated layers by crustacea and worms. In addition, physical stirring via sediment "bursting" (14, 15) can facilitate exchange across the benthic interface. In general, transport across the benthic boundary layer can be described via an advection/dispersion equation (16), in which the advection term accounts for ground-water flow and exchange events are described via a dispersion term.

Biosorption and Bioaccumulation. Food chain transmission of pollutants, and direct pollutant transfer to otherwise uncontaminated populations (e.g., human consumption of contaminated fishes), result from biosorption and bioaccumulation of xenobiotic chemicals. In the case of microorganisms, with their very high surface to volume ratio, capture of synthetic organics can be described via a partition coefficient (17). Simple partitioning probably can be used to describe uptake of synthetic organics by the majority of organisms occupying the base of aquatic food chains. The uptake of pollutant chemicals by higher organisms is complicated by the presence of multiple dietary and direct routes of exposure, including transport across gill membranes and, at least for marine fishes, direct water intake. The significance of partitioning, as against dietary uptake, must vary with the physiology and life history of the organism. For example, blue crabs (Callinectes) take up little or no Kepone via direct absorption from seawater, but readily accumulate this compound via dietary intake (18). Although simple bioconcentration factors are adequate for many purposes, detailed exposure models probably should include some capability for representing dietary exposures (19), biosorption or membrane transfers, and depuration, metabolism, and detoxification.

Volatilization. Transfer of chemicals across the air/water interface can result in either a net gain or loss of chemical, although in many cases the bulk concentration in the air above a contaminated water body is low enough to be neglected (20). When the atmosphere is the primary source of the contaminant, as for example polychlorinated biphenyls in some parts of the Laurentian Great Lakes, atmospheric concentrations obviously cannot be neglected. The Whitman two-film or two-resistance approach (21) has been applied to a number of environmental situations (20, 22, 23). Transport across the air/water interface is viewed as a two-stage process, in which both phases of the interface can offer resistance to transport of the chemical. The rate of transfer depends on turbulence in the water body and in the atmosphere, the

Henry's Law constant of the chemical, and the molecular velocity
of the compound in the near-surface regions.

Transformation Processes

Direct Photolysis. Direct photochemical reactions are due to
absorption of electromagnetic energy by a pollutant. In this
"primary" photochemical process, absorption of a photon promotes a
molecule from its ground state to an electronically excited state.
The excited molecule then either reacts to yield a photoproduct or
decays (via fluorescence, phosphorescence, etc.) to its ground
state. The efficiency of each of these energy conversion
processes is called its "quantum yield"; the law of conservation
of energy requires that the primary quantum efficiencies sum to
1.0.   Photochemical reactivity is thus composed of two factors:
the absorption spectrum, and the quantum efficiency for
photochemical transformations.
     The rate of photolytic transformations in aquatic systems also
depends on the intensity and spectral distribution of light in the
medium (24). Light intensity decreases exponentially with depth.
This fact, known as the Beer-Lambert law, can be stated
mathematically as: $d(Eo)/dZ = -K(Eo)$, where Eo = photon scalar
irradiance (photons/cm^2/sec), Z = depth (m), and K = diffuse
attenuation coefficient for irradiance (/m). The product of light
intensity, chemical absorptivity, and reaction quantum yield, when
integrated across the solar spectrum, yields a pseudo-first-order
photochemical transformation rate constant.

Indirect Photolysis and Oxidation. The simultaneous occurrence of
"humic materials," dissolved oxygen, and sunlight often results in
an acceleration of the rate of transformation of organic
pollutants (25, 26). These reactions are termed "indirect" or
"sensitized" photolysis. Indirect photolysis can be subdivided
into two classes of reactions. First, "sensitized photolysis" per
se involves excitation of a (humic) sensitizer by sunlight,
followed by direct chemical interaction between the sensitizer and
a pollutant. The second class of indirect photolysis involves the
formation of chemical oxidants, primarily via the interaction of
sunlight, humic materials, and dissolved oxygen (27). The primary
oxidants known to occur in natural waters are hydroxyl and peroxy
radicals (28) and singlet oxygen (29). Aquatic fate codes
represent the oxidative transformation of pollutants via a purely
phenomenological coupling of a second-order rate constant (with
units /M/time) to the concentration (moles/liter) of oxidants in
the system. The occurrence and concentration of oxidants in
natural waters has been investigated via the laboratory
irradiation of chemical "probes" (28, 29, 30).

Hydrolysis. Hydrolysis of synthetic organics occurs via several
pathways. Specific-acid and -base catalyzed processes can be

described in terms of pH and temperature (31). Neutral hydrolysis
(i.e., pH-independent mechanisms) can proceed either via direct
reactions involving the water molecule in the rate-limiting step
or via indirect molecular transformations involving water as a
reactant. (Both are apparent first-order reactions.) The
information required by fate models can be encapsulated in a
pH-rate profile (Figure 1). Hydrolysis kinetics are readily
described via second-order equations incorporating the effects of
temperature and ionization on chemical reactivity.

General acid/base catalysis is less significant in natural
fresh waters, although probably of some importance in special
situations. This phenomenon can be described fairly well via the
Bronsted law (relating rate constants to pKa and/or pKb of general
acids and bases). Maximum rates of general acid/base catalysis
can be deduced from a compound's specific acid/base hydrolysis
behavior, and actual rates can be determined from relatively
simple laboratory experiments (34).

<u>Microbial Biotransformation</u>. Microbial population growth and
substrate utilization can be described via Monod's (35) analogy
with Michaelis-Menten enzyme kinetics (36). The growth of a
microbial population in an unlimiting environment is described by:
dN/dt = u N, where u is called the "specific growth rate" and N is
microbial biomass or population size. The Monod equation modifies
this by recognizing that consumption of resources in a finite
environment must at some point curtail the rate of increase
(dN/dt) of the population:

$$u = u(max) \frac{[S]}{Ks + [S]} \qquad (2)$$

in which [S] is the concentration of the growth-limiting
substrate, u(max) is the "maximum specific growth rate" attained
when [S] is present in excess (i.e., non-limiting), and Ks, the
"saturation constant" is that value of [S] allowing the population
to grow at rate u(max)/2. An equation describing the behavior of
[S] over time, and thus by implication the dynamics of a
biodegradable synthetic compound, follows via a simple derivation
(36), resulting in Equation 3, in which Y is the "yield
coefficient" in cells or biomass produced per unit S metabolized.

$$\frac{dS}{dt} = - \frac{u(max)}{Y} \frac{[S]}{Ks + [S]} N \qquad (3)$$

This equation is difficult to apply in a broader ecological
context, however. The first difficulty is mechanical: Equation 3
is non-linear in its parameters, and thus imposes a high cost in
computation time when used in a computer code. For trace

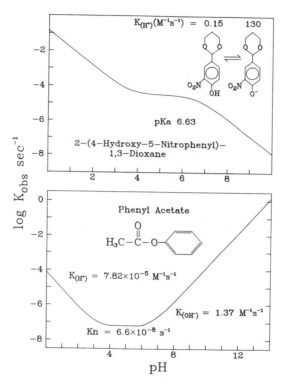

Figure 1.   Hydrolysis pH–rate profiles of phenyl acetate (lower) and a substituted 2–phenyl–1,3–dioxane (HND). Phenyl acetate profile constructed from data of Mabey and Mill (32), HND profile from data of Bender and Silver (33).   Phenyl acetate reacts via specific–acid catalyzed, neutral, and base–catalyzed transformation pathways.   The pseudo–first–order rate constant is given by Kobs = K$_{(H+)}$ [H$^+$] + Kn + K$_{(OH-)}$ [OH$^-$].   HND hydrolyzes only via an acid–catalyzed pathway; the phenolate anion is some 867 times more reactive than its conjugate acid. In this case, Kobs = (0.15 Io + 130(1–Io))[H$^+$] (See equation 1.)

concentrations (i.e., [S] << Ks), however, the term (Ks + [S])   in
Equation 3 can be approximated by Ks, giving:

$$\frac{d[S]}{dt} = - \frac{u(max)}{Y \, Ks} N \, [S] \qquad (4)$$

This formulation is similar to the "second-order" equations used
to describe the kinetics of chemical reactions, and u(max)/(Y Ks)
can by analogy be termed a second-order biolysis rate constant
Kb2, with units /time/(cells/liter) when population sizes N are
expressed in cells/liter.

There are conceptual difficulties as well, however. A natural
microbial community derives its energy from a variety of sources.
A species restricted to a trace-level synthetic compound as its
sole carbon source would be at a severe competitive disadvantage;
there is no way to predict the population densities the degrader
could attain in real systems. Further, the presence of multiple
energy-yielding substrates violates a fundamental assumption of
the Monod approach, that is, that [S], the synthetic compound,
limits growth. Compounds may be transformed in energy-requiring
detoxification processes, making the concept of cell yield (Y) of
dubious utility. When the compound is degraded without a   change
in population size, the zero yield invalidates Equations 3 and 4
alike. This phenomenon is sometimes called "cometabolism" (37).

Still, the rate of transformation of organic pollutants must
depend on the structure of the compound and the metabolic capacity
of microbial communities. The simplest expression of this duality
is the second-order equation:

$$d[S]/dt = - Kb2 \, B \, [S] \qquad (5)$$

which asserts that the rate of biolysis (d[S]/dt) is first-order
in compound concentration [S] and in microbial activity B, and in
which a dogmatic identification of Kb2 with u(max)/(Y Ks) is
discarded.   This technique is used by most currently operational
exposure models.

Logical and Data Structures of Aquatic Fate Codes

Logical Structures. When a synthetic organic chemical is released
into an aquatic system, the entire array of transport, transfer,
and transformation processes begins at once to act on the
chemical.   Transport from the point of entry into the bulk of the
system takes place by advection and by turbulent dispersion.
Transfers to sorbed forms and irreversible transformation
processes proceed simultaneously with the transport of the
chemical. After the elapse of sufficient time, the chemical comes
to be distributed throughout the system, with relatively smooth
concentration gradients resulting from dilution, speciation, and

transformation. The most efficient way to describe the parallel action of the processes is to combine them into a mathematical description of their total effect on the rate of change of chemical concentration in the system, via a set of coupled differential equations.

The simplest, and perhaps most important, principle used in constructing environmental models is the conservation law: matter can be neither created nor destroyed. This means that every molecule of a chemical that enters a defined spatial zone (e.g., a laboratory tank or beaker, the epilimnion of a lake, the Earth's atmosphere, etc.) must ultimately either leave the system, be transformed into another compound, or take up residence in the system. The behavior of a chemical in an aquatic system thus can be rigorously described by using the conservation law as an accounting principle. Imagine, first, an accounting boundary drawn around some segment of the environment. This imaginary boundary encloses a "control volume," or accounting unit to be used in a model. In real cases, accounting boundaries are often chosen to correspond with actual physical discontinuities, such as river banks, the air/water interface, the benthic-sediment/water-column interface, the depth of bioturbation of sediments, etc. In many cases, however, the accuracy and detail of a model can be improved by including many relatively small accounting units bounded only by arbitrary lines drawn on a map or vertical profile of the water body.

Every molecule entering the control volume has three courses available to it: export (movement out of the control volume), transformation, or residence. Environmental modeling thus begins by writing a mass balance around the control volume. This mass balance expresses the changes in concentration (that is, increases and decreases in the number of chemical molecules resident per unit volume of the segment) that result from loadings (i.e., inputs), exports, and transformations. The product of concentration [C] (mass/volume) and volume V is the accountable mass, so, using the usual notation of differential equations to denote a rate of change (that is, $d[C]/dt$ denotes the rate of change of concentration [C] per unit time t):

$$V \, d[C]/dt = Lo - Ex - V k E [C] \qquad (6)$$

that is, mass change rate = load - export - transformations, where E is the environmental factor driving the transformation of the compound (e.g., $[H^+]$), and k is the kinetic constant of the reaction pathway. If we consider water-borne exports only, Ex can be set equal to (F)[C], where (F) is the flow rate of water leaving the system in volume/time. The second-order transformation process k E [C] must have units mass/volume/time; multiplication by V converts it to the proper units (mass/time).

Dividing both sides of Equation 6 by V gives a unit equation for modeling chemical concentrations in real systems:

$$d[C]/dt = Lo/V - (F/V)[C] - k E [C] \qquad (7)$$

(The term (V/F) has units of time; when V and F apply to an entire system, (V/F) is often called the "hydraulic residence time" or "detention time" of the water body.)

Models of real ecosystems often consist of many unit control volumes coupled together by transport equations. An export from one unit then becomes a load on another. The loadings (Lo) on each control volume can thus be the sum of many terms, including "external" chemical loadings from industrial or agricultural discharges, contaminated rainfall, etc., plus the "internal" loadings due to hydrodynamic motions and other transport cycles within the system that move chemicals from one control volume to another. In models that include both parent compounds and transformation products, the rate of transformation of the parent to the product (k E [C]) is also an internal loading (Lo/V) of the product chemical.

Export processes are often more complicated than the expression given in Equation 7, for many chemicals can escape across the air/water interface (volatilize) or, in rapidly depositing environments, be buried for indeterminate periods in deep sediment beds. Still, the majority of environmental models are simply variations on the mass-balance theme expressed by Equation 7. Some codes solve Equation 7 directly for relatively large control volumes, that is, they operate on "compartment" or "box" models of the environment. Models of aquatic systems can also be phrased in terms of continuous space, as opposed to the "compartment" approach of discrete spatial zones. In this case, the partial differential equations (which arise, for example, by taking the limit of Equation 7 as the control volume goes to zero) can be solved by finite difference or finite element numerical integration techniques.

Data Structures. Inspection of the unit simulation equation (Equation 7) indicates the kinds of input data required by aquatic fate codes. These data can be classified as chemical, environmental, and loading data sets. The chemical data set, which are composed of the chemical reactivity and speciation data, can be developed from laboratory investigations. The environmental data, representing the driving forces that constrain the expression of chemical properties in real systems, can be obtained from site-specific limnological field investigations or as summary data sets developed from literature surveys. Allochthonous chemical loadings can be developed as worst-case estimates, via the outputs of terrestrial models, or, when appropriate, via direct field measurement.

From these data, aquatic fate models construct outputs
delineating exposure, fate, and persistence of the compound. In
general, exposure can be determined as a time-course of chemical
concentrations, as ultimate (steady-state) concentration distributions, or as statistical summaries of computed time-series.
Fate of chemicals may mean either the distribution of the chemical
among subsystems (e.g., fraction captured by benthic sediments),
or a fractionation among transformation processes. The latter
data can be used in sensitivity analyses to determine relative
needs for accuracy and precision in chemical measurements.
Persistence of the compound can be estimated from the time
constants of the response of the system to chemical loadings.

Available Aquatic Fate Codes

Although many research models and water quality codes have been
written in recent years, relatively few finished codes for
synthetic organics in aquatic systems are available to
non-specialists. Useful codes must include some form of the
second-order process models described above, be written as general
purpose codes (i.e., not as regression models), have some form of
written documentation and user manual, and be publicly available
in a portable high-level language. All the codes listed below
survive these tests. All are written in FORTRAN IV; FORTRAN
compilers are virtually universally available. A brief
description of the main available codes follows. The references
given in these paragraphs include documentation and sources for
additional information and copies of the computer programs. These
codes are, for the most part, undergoing continuing revision and
test by their authors, so detailed comparisons are problematic and
are not attempted. The documentation reports cited for the codes
may not fully describe the currently available programs; direct
contact with code authors is advisable.

A code developed for the Chemical Manufacturers Association by
HydroQual (SLSA, or Simplified Lake and Stream Analyzer **38**, **39**)
deserves careful study for its insightful exploration of the
general characteristics of the behavior of hydrophobic materials
in aquatic systems. This code, with its documentation report,
provides an excellent entry point to the field.

EXAMS1/EXAMS2. The EXposure Analysis Modeling System (EXAMS, **40**)
was developed as a screening tool for estimating long-term fate,
residual concentrations, and persistence. Although written in
full differential equation form, the program invokes a special
routine for deriving steady-state solutions, so as to simplify
evaluation of long-term chemical loadings. EXAMS can accept one,
two, or three-dimensional compartment models of aquatic systems.
A follow-on version (EXAMS2) computes the fate of transformation
products and allows direct user access to initial conditions and
simulation intervals.

HSPF. The Hydrologic Simulation Program (FORTRAN) (<u>41</u>, <u>42</u>) is based on the Stanford Watershed Model. Version 7 of HSPF incorporates the process models of SERATRA in its aquatic section, with several (user-selectable) options for sediment transport computations. HSPF includes the generation of transformation products, each of which is in turn subject to volatilization, phototransformation, biolysis, etc.

PEST. This code (<u>43</u>) was developed within the framework of Rensselaer Polytechnic Institute's CLEAN (Comprehensive Lake Ecosystem Analyzer) model. It includes highly elaborated algorithms for biological phenomena, as described in this volume (<u>44</u>). For example, biotransformation is represented via second-order equations in bacterial population density (Equation 5) in the other codes described in this section; PEST adds to this effects of pH and dissolved oxygen on bacterial activity, plus equations for metabolism in higher organisms. PEST allows for up to 16 compartments (plants, animals, etc.), but does not include any spatially resolved computations or transport processes other than volatilization.

SRI/SERATRA/TODAM/FETRA/CMRA. Based on a summary of process models published by Stanford Research Institute (SRI model, <u>45</u>), Battelle Pacific Northwest Laboratories' SERATRA (Sediment-Radionuclide Transport Model) was expanded from its original first-order radionuclide decay to include process models for photolysis, hydrolysis, oxidation, biolysis, and volatilization (<u>46</u>). The SERATRA code is an unsteady, two dimensional (longitudinal and vertical) finite element code for transport of dissolved constituents and transport, deposition, and resuspension of sediments and sorbed contaminants. A one-dimensional (longitudinal) version (TODAM) and a two-dimensional (longitudinal and lateral) estuarine version (FETRA) are also available from BPNL. These codes require hydrodynamic conditions over time as input to the programs. Sorption/desorption are treated as kinetic processes first-order in chemical concentrations. A set of codes directly useful in environmental safety evaluations (CMRA, Chemical Migration and Risk Assessment) was also developed for use with the SERATRA program (<u>47</u>). CMRA also includes a non-point-source loading model for agricultural sources, a hydrodynamic code, and a toxicological post-processor FRANCO (FRequency ANalysis of pesticide COncentrations for risk assessment) that assembles time-series data on chemical concentrations into an analysis of the frequency and duration of toxic stress events.

TOXIC. This code (<u>48-51</u>) was developed at the University of Iowa as an elaboration of the SRI model. To its predecessor, it adds a fish uptake and depuration model and expanded dynamic capabilities. The code was developed during the course of field studies of the behavior of pesticides and herbicides in Iowa reservoirs.

UTM-TOX. The Unified Transport Model for Toxicants (UTM-TOX, 52) was developed at Oak Ridge National Laboratory (ORNL) on the base of the ORNL Unified Transport Model (UTM), itself under development from the early 1970's. UTM-TOX includes air, water, and terrestrial submodels. The aquatic fate submodel includes volatilization, hydrolysis, biolysis, photolysis, and sorption equilibria. Sorbed phases are assumed unreactive in the 1982 version.

WASP/TOXIWASP/WASTOX. The Water Quality Analysis Simulation Program (WASP, 53) is a generalized finite-difference code designed to accept user-specified kinetic models as subroutines. It can be applied to one, two, and three-dimensional descriptions of water bodies, and process models can be structured to include linear and non-linear kinetics. Two versions of WASP designed specifically for synthetic organic chemicals exist at this time. TOXIWASP (54) was developed at the Athens Environmental Research Laboratory of U.S. E.P.A.; WASTOX (55) was developed at HydroQual, with participation from the group responsible for WASP. Both codes include process models for hydrolysis, biolysis, oxidations, volatilization, and photolysis. Both treat sorption/desorption as local equilibria. These codes allow the user to specify either constant or time-variable transport and reaction processes.

All these codes vary slightly in implementation, resulting in slightly different input data. For example, some use standard Arrhenius functions to express temperature dependencies; others use the civil engineering power equation. Although equivalent, the input data differ. Again, some codes use the 39 wavelength intervals specified by the SOLAR photochemical code (24); others use an 18 increment version from SERATRA. A few codes do not include a term for atmospheric resistance to volatilization, and thus require caution when used for compounds with Henry's Law constants smaller than 0.001 atm-m^3/mole (23). At present, only HSPF and EXAMS2 include transformation product capabilities, and only the EXAMS codes include a full treatment of ionizable compounds and ion-specific chemical reactivities. It should be said, however, that most of the currently active code authors are aware of the limitations of their codes and are actively engaged in further development.

Although all these codes have been tested for theoretical integrity and realism to some extent, few appropriate data sets for model testing are available and standardized methods for collecting them are only now under development (56). At this time, a full evaluation of a chemical is perhaps best accomplished using several fate codes, with careful comparisons among the outputs of the codes.

Literature Cited

1. Baughman, G.L.; Burns, L.A. in "The Handbook of Environmental Chemistry, Vol. 2 Part A: Reactions and Processes"; Hutzinger, O., Ed.; Springer-Verlag: Berlin, 1980; pp. 1-17.
2. Karickhoff, S.W. in "Contaminants and Sediments, Vol. 2: Analysis, Chemistry, Biology"; Baker, R.A., Ed.; Ann Arbor Science Publ.: Ann Arbor, Michigan, 1980; pp. 193-205.
3. Mackay, D. Environ. Sci. Technol. 1979, 13, 1218-1223.
4. McCall, P.J.; Swann, R.L.; Laskowski, D.A. Chapter __ in this book.
5. Shen, H.W., Ed.; "Modeling of Rivers"; Wiley: New York, 1979.
6. Nihoul, J.C.J., Ed.; "Hydrodynamics of Estuaries and Fjords"; Elsevier Scientific Publishing Co.: Amsterdam, 1978; p. 546.
7. Gibbs, R.J., Ed.; "Transport Processes in Lakes and Oceans"; Plenum: New York, 1977; p. 288.
8. Stumm, W.; Morgan, J.J. "Aquatic Chemistry"; Wiley: New York, 1970; p. 583.
9. Karickhoff, S.W.; Brown, D.S.; Scott, T.A. Water Res. 1979, 13, 241-248.
10. Brown, D.S., personal communication.
11. Garde, R.J.; Ranga Raju, K.G. "Mechanics of Sediment Transportation and Alluvial Stream Problems"; Wiley: New York, 1977; p. 483.
12. Simons, D.B.; Senturk, F. "Sediment Transport Technology"; Water Resources Publ.: Fort Collins, Colorado, 1977; p. 807.
13. Aller, R.C. Ph.D. Thesis, Yale University, New Haven, Connecticut, 1977.
14. Heathershaw, A.D. Nature 1974, 248, 394-395.
15. Heathershaw, A.D. in "The Benthic Boundary Layer"; McCave, I.N., Ed.; Plenum: New York, 1976; pp. 11-31.
16. Berner, R.A. "Early Diagenesis: A Theoretical Approach"; Princeton Univ. Press: Princeton, N.J., 1980; p. 241.
17. Baughman, G.L.; Paris, D.F. CRC Critical Reviews Microbiol., 1981, 8, 205-228.
18. Schimmel, S.C.; Patrick, J.M. Jr.; Faas, L.F.; Oglesby, J.L.; Wilson, A.J. Jr. Estuaries, 1979, 2, 9-15.
19. Weininger, D. Ph.D. Thesis, Univerisity of Wisconsin, Madison, 1978.
20. Mackay, D.; Leinonen, P.J. Environ. Sci. Technol., 1975, 9, 1178-1180.
21. Whitman, R.G. Chem. Metallurg. Eng., 1923, 29, 146-148.
22. Liss, P.S.; Slater, P.G. Nature, 1974, 247, 181-184.
23. Mackay, D. in "Aquatic Pollutants: Transformation and Biological Effects"; Hutzinger, O.; van Lelyveld, I.H.; Zoeteman, B.C.J., Eds.; Pergamon: Oxford, 1978; pp. 175-185.
24. Zepp, R.G.; Cline, D.M. Environ. Sci. Technol., 1977, 11, 359-366.
25. Zepp, R.G.; Wolfe, N.L.; Azarraga, L.V.; Cox, R.H.; Pape, C.W. Arch. Environm. Contam. Toxicol., 1977, 6, 305-314.

26. Ross, R.D.; Crosby, D.G. Chemosphere, 1975, 4, 277-282.
27. Zepp, R.G.; Baughman, G.L. in "Aquatic Pollutants: Transformation and Biological Effects"; Hutzinger, O.; van Lelyveld, I.H.; Zoeteman, B.C.J., Eds.; Pergamon: Oxford, 1978; pp. 237-263.
28. Mill, T.; Hendry, D.G.; Richardson, H. Science, 1980, 207, 886-887.
29. Zepp, R.G.; Wolfe, N.L.; Baughman, G.L.; Hollis, R.C. Nature, 1977, 267, 421-423.
30. Mill, T.; Richardson, H.; Hendry, D.G. in "Aquatic Pollutants: Transformation and Biological Effects"; Hutzinger, O.; van Lelyveld, I.H.; Zoeteman, B.C.J., Eds.; Pergamon: Oxford, 1978; pp, 223-236.
31. Wolfe; N.L. in "Dynamics, Exposure and Hazard Assessment of Toxic Chemicals"; Haque, R., Ed.; Ann Arbor Science Publ.: Ann Arbor, Michigan, 1980; pp. 163-178.
32. Mabey, W.; Mill, T. J. Phys. Chem. Ref. Data, 1978, 7, 383-415.
33. Bender, M.L.; Silver, M.S. J. Amer. Chem. Soc., 1963, 85, 3006-3010.
34. Perdue, E.M.; Wolfe, N.L. Environ. Sci. Technol., 1983; in press
35. Monod, J. "Recherches sur la Croissance des Cultures Bacteriennes"; Herman: Paris, 1942.
36. Slater, J.H. in "Microbial Ecology: A Conceptual Approach"; Lynch, J.M.; Poole, N.J.; Eds.; Blackwell: Oxford, 1979; pp. 45-63.
37. Alexander, M. in "Dynamics, Exposure and Hazard Assessment of Toxic Chemicals"; Haque, R., Ed.; Ann Arbor Science Publ.: Ann Arbor, Michigan, 1980; pp. 179-190.
38. Di Toro, D.M.; O'Connor, D.J.; Thomann, R.V.; St. John, J.P.; "Analysis of Fate of Chemicals in Receiving Waters--Phase I"; (CMA Project ENV-7-W); HydroQual: Mahwah, New Jersey, 1981.
39. Di Toro, D.M.; O'Connor, D.J.; Thomann, R.V.; St. John, J.P. in "Modeling the Fate of Chemicals in the Aquatic Environment"; Dickson, K.L.; Cairns, J., Jr.; Maki, A.W., Eds.; Ann Arbor Science Publ.: Ann Arbor, Michigan, 1982; pp. 165-190.
40. Burns, L.A.; Cline, D.M.; Lassiter, R.R. "EXposure Analysis Modeling System (EXAMS): User Manual and System Documentation"; EPA-600/3-82-023, U.S. Environ. Prot. Agency, Environ. Research Lab.: Athens, Georgia, 1982; p. 443.
41. Johanson, R.C.; Chapter ___ in this book.
42. Johanson, R.C.; Imhoff, J.C.; Davis, H.H., Jr. "Users Manual for Hydrological Simulation Program - FORTRAN (HSPF)"; EPA-600/9-80-015, U.S. Environ. Prot. Agency, Environ. Research Lab.: Athens, Georgia, 1980; p. 678.
43. Park, R.A.; Connolly, C.I.; Albanese, J.R.; Clesceri, L.S.; Heitzman, G.W.; Herbrandson, H.H.; Indyke, B.H.; Lohe, J.R.; Ross, S.; Sharma, D.D.; Shuster, W.W. "Modeling the Fate of

Toxic Organic Materials in Aquatic Environments"; EPA-600/3-82-028, U.S. Environ. Prot. Agency, Environ. Research Lab.: Athens, Georgia, 1982; p. 163.

44. Park, R.A.; Clesceri, L.S.; Chapter ___ in this book.

45. Smith, J.H.; Mabey, W.R.; Bohonos, N.; Holt, B.R.; Lee, S.S.; Chou, T.-W.; Bomberger, D.C.; Mill, T. "Environmental Pathways of Selected Chemicals in Freshwater Systems. Part I: Background and Experimental Procedures"; EPA-600/7-77-113, U.S. Environ. Prot. Agency, Environ. Research Lab.: Athens, Georgia, 1977; p. 80.

46. Onishi, Y.; Wise, S.E. "Mathematical Model, SERATRA, for Sediment - Contaminant Transport in Rivers and its Application to Pesticide Transport in Four Mile and Wolf Creeks in Iowa"; EPA-600/3-82-045, U.S. Environ. Prot. Agency, Environ. Research Lab.: Athens, Georgia,1982; p. 56.

47. Onishi, Y.; Brown, S.M.; Olsen, A.R.; Parkhurst, M.A.; Wise, S.E.; Walters, W.H. "Methodology for Overland and Instream Migration and Risk Assessment of Pesticides"; EPA-600/3-82-024, U.S. Environ. Prot. Agency, Environ. Research Lab.: Athens, Georgia, 1982; p. 115 + Appendices.

48. Schnoor, J.L.; Rao, N.; Cartwright, K.J.; Noll, R.M.; Ruiz-Calzada, C.E. "Verification of a Toxic Organic Substance Transport and Bioaccumulation Model"; EPA-600/3-83-007, U.S. Environ. Prot. Agency, Environ. Research Lab.: Athens, Georgia, 1983; p. 164.

49. Schnoor, J.L. Science 1981, 211, 840-842.

50. Schnoor, J.L.; McAvoy, D.C. J. Environ. Engr. Div., ASCE 1981, 107(EE6), 1229-1246.

51. Schnoor, J.L. Chapter ___ in this book.

52. Browman, M.G.; Patterson, M.R.; Sworski, T.J. "Formulations of the Physicochemical Processes in the ORNL Unified Transport Model for Toxicants (UTM-TOX) Interim Report"; ORNL/TM-8013, Oak Ridge Nat. Laboratory, Oak Ridge, Tennessee, 1982; p. 46.

53. Di Toro, D.M.; Fitzpatrick, J.J.; Thomann, R.V. "Water Quality Analysis Simulation Program (WASP) and Model Verification Program (MVP)--Documentation"; U.S. Environ. Prot. Agency, Environ. Research Lab.: Duluth, Minnesota, Contract No. 68-01-3872, 1981; p. 135.

54. Ambrose, R.B., Jr.; Hill, S.I.; Mulkey, L.A. "User's Manual for the Toxic Chemical Transport and Fate Model (TOXIWASP) Version I"; EPA-600/3-83-005 U.S. Environ. Prot. Agency, Environ. Research Lab.: Athens, Georgia, 1983; p. 95.

55. Connolly, J.P. "WASTOX Preliminary Estuary and Stream Version Documentation"; U.S. Environ. Prot. Agency, Environ. Research Lab.: Gulf Breeze, Florida, 1982; p. 96 (draft).

56. Crockett, A.B.; Hern, S.C.; Kinney, W.L.; Flatman, G.T. "Guidelines for Field Testing Aquatic Fate and Transport Models: Interim Report"; U.S. Environ. Prot. Agency, Environ. Monitoring Systems Lab.: Las Vegas, Nevada, 1982; p. 174 + Appendices.

RECEIVED April 15, 1983.

Soil and Groundwater Fate Modeling

MARCOS BONAZOUNTAS

Arthur D. Little, Inc., Cambridge, MA 02140

Soil compartment chemical fate modeling has been traditionally performed for three distinct subcompartments: the land surface (or watershed);
the unsaturated soil (or soil) zone; and the saturated (or groundwater) zone
of a region. In general, the mathematical simulation is structured around
two major cycles: the hydrologic cycle and the pollutant cycle, each cycle
being associated with a number of physicochemical processes. Watershed
models account for a third cycle: sedimentation.

This paper discusses: (1) soil and groundwater; and (2) aquatic equilibrium and ranking models. The second category deals with the chemical
"speciation" in soil and groundwater, and with the environmental rating of
waste sites, in cases where detailed modeling is not desirable.

Unsaturated soil zone models can simulate flow and quality conditions
of a soil zone profile extending between the ground surface and the
groundwater table. The traditional modeling employs either the time
dependent diffusive convective mass transport differential equation in
homogeneous isotropic soils, or simplified analytic expressions. Compartmental modeling is an area of current research. Equations are principally
solved numerically or analytically when seeking exact solutions for simplified environments. Groundwater models describe the fate of pollutants in
aquifers. Ironically enough, although the number of model types is large,
only a few basic processes are modeled, mainly via the convective, dispersive, adsorptive, reactive, pollutant transport equation in a saturated
porous medium, which is solved via analytical, numerical, or statistical
techniques, and for one pollutant at a time.

1.0 INTRODUCTION

Soil compartment chemical fate modeling has been traditionally performed for
three distinct subcompartments: (1) land surface (or watershed); (2) the unsaturated
soil (or soil) zone, and (3) the saturated (or groundwater) zone of a region (Figure 1). In
general, the mathematical simulation is structured around two major cycles: the
hydrologic cycle and the pollutant cycle, each cycle being associated with a number of
physicochemical processes. Land surface models account for a third cycle: sedimentation. Land surface models describe pollutant fate on the watershed and pollutant
contribution to a water body and to the unsaturated soil zone of the region. Unsaturated soil zone models simulate flow and quality conditions of a soil zone profile

0097–6156/83/0225–0041$07.25/0

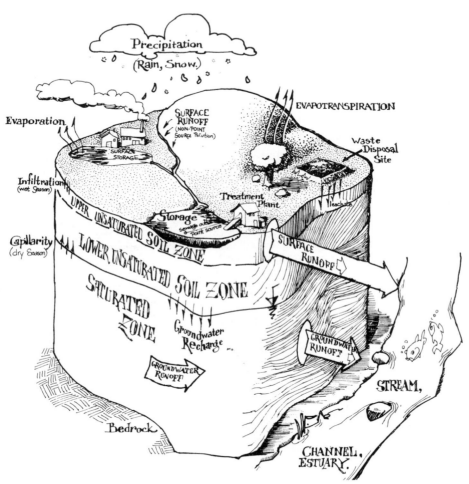

Figure 1. Schematic presentation of the soil compartment.
(Reproduced with permission from Ref. 5.)

extending between the ground surface and the groundwater table. Groundwater models describe the fate of pollutants in aquifers.

Mathematical models can greatly assist decision makers in determining the importance of pollutant pathways in the environment, as long as used properly and with an understanding of their limitation. The use of models has grown dramatically over the past decade, but models are not meant to substitute for good judgment and experience. Ironically enough, although the number of model types is very large, only a few basic different modeling concepts exist.

Soil zone modeling is a very complex issue and a major characteristic of a soil subcompartment — as contrasted to a water or an air subcompartment — is that the temporal physical and the chemical behavior of this subcompartment is governed by "out-compartmental" forces such as precipitation, air temperature, and solar radiation. This characteristic is also one of the main reasons why soil mathematical modeling can be much more complex than water or air modeling. Groundwater modeling can handle a limited number of chemical processes; therefore, a number of aquatic equilibrium models under development are aimed to fill this chemistry gap. Finally, soil compartment models (watershed, soil, groundwater) are used to evaluate pollution originating from various sources, such as hazardous waste sites. Because detailed soil modeling is not always feasible or desirable, a category of "ranking" models is known in the literature for "screening" severity of environmental impacts originating from waste activities.

This paper presents a review discussion of soil, groundwater, aquatic equilibrium and ranking modeling concepts including selected documented models. Watershed models are not discussed, since the work of Knisel ($\underline{1}$) is one of the most representative watershed computerized packages.

2.0 SOURCES AND EMISSIONS

Through numerous human activities pollutants are released to the soil compartment. The particular practice significantly influences the fate of pollutants in the soil and groundwater zones. Releases include both point source and area loadings. They may be intentional, such as landfills and spray irrigation of sewage; unintentional, such as spills and leaks; or indirect, through pesticide drift or surface runoffs. The point of release may be at the soil surface, or from a source buried deep in the soil. Substances released are in liquid, semi-liquid, solid, or particulate form. In some cases a waste material will be pretreated or deactivated prior to disposal to limit its mobility in soil. The rate of release may be continuous, such as at a municipal landfill; intermittent, or on a "batch" basis such as practiced by some industries, or as a one-time episode such as the uncontrolled disposal of barrels of hazardous waste or a spill.

Table 1 indicates primary pollutant sources and waste modes, and Table 2 indicates the primary and secondary sources and associated pollutants. The primary sources of soil contamination include: land disposal of solid waste; sludge and wastewater; industrial activities; and leakages and spills, primarily of petroleum products. The solid waste disposal sites include: dumps, landfills, sanitary landfills, and secured landfills.

Land disposal sites result in soil contamination through leachate migration. The composition of the substances produced depends principally on the type of wastes present and the decomposition in the landfill (aerobic or anaerobic). The adjacent soil can be contaminated by: direct horizontal leaching of surface runoff; vertical leaching; and transfer of gases from decomposition by diffusion and convection. The disposal of

TABLE 1

SOURCES AND WASTES

Pollutant Source	Wastewater Impoundments	Solid Waste Disposal Sites	Wastewater-Spray Irrigation	Land Application	Injection or Disposal Wells	Septic Tanks and Cesspools	Infiltration/Surface Runoff — Pits	Infiltration/Surface Runoff — Surface Runoff	Leaching from Storage Sites
Industrial									
— Wastewater	X		X		X				
— Sludge				X					
— Solid Waste		X							X
Municipal									
— Wastewater			X						
— Sludge		X		X					
— Solid Waste		X							X
Household									
— Wastewater						X			
Agricultural Feedlot								X	
Mining	X	X			X		X	X	
Petroleum Exploration					X		X		
Cooling Water					X				
Buried Tanks, Pipelines									X
Agricultural Activities								X	X

TABLE 2

PRIMARY SOURCES OF SOIL CONTAMINATION AND ASSOCIATED POLLUTANTS

Source	Type of Pollutants
Industrial Sources	
Chemical manufacturers	Organic solvents
Petroleum refineries	Chlorinated hydrocarbons
Metal smelters and refineries	Heavy metals
Electroplaters	Cyanide, other toxics
Paint, battery manufacturers	Conventional pollutants
Pharmaceutical manufacturers paper and related industries	Acids, alkalines, other corrosives many are highly mobile in soil.
Land Disposal Sites	
Landfills that received sewage sludge, garbage, street refuse, construction and demolition wastes	BOD, inorganic salts, heavy metals pathogens, refractory organic compounds, plastics; nitrate; metals including iron, copper, manganese suspended solids
Uncontrolled dumping of industrial wastes, hazardous wastes	
Mining Wastes	Acidity, dissolved solids, metals, radioactive materials, color, turbidity
Agricultural Activities	BOD, nutrients, fecal coliforms, chloride, some heavy metals
Agricultural feedlots	
Treatment of crops and/or soil with pesticides and fertilizers; runoff or direct vertical leaching to septic tanks and cesspools	Herbicides, insecticides, fungicides, nitrates, phosphates, potassium, BOD, nutrients, heavy metals, inorganic salts, pathogens, surfactants; organic solvents used in cleaning
Leaks and Spills	
Sources include oil and gas wells; buried pipelines and storage tanks; transport vehicles	Petroleum and derivative compounds; any transported chemicals
Atmospheric deposition	Particulates; heavy metals, volatile organic compounds; pesticides; radioactive particles
Highway Maintenance Activities	
Storage areas and direct application	Primarily salts
Radioactive Waste Disposal	
Eleven major shallow burial sites exist in U.S.; 3 known to be leaking	Primarily ^{132}CS, ^{90}Sr, and ^{60}Co
Land Disposal of Sewage and Wastewater	
Spray irrigation of primary, secondary effluents	BOD, heavy metals, inorganic salts, pathogens, nitrates, phosphates, recalcitrant organics
Land application of sewage sludge	
Leakage from sewage oxidation ponds	

domestic and municipal wastewaters on land takes place through: septic tanks and cesspools; sewage sludge from primary and secondary treatment plants often spread on agricultural and forested land (land treatment); liquid sewage, either untreated or partially treated is applied to the land surface, by spray irrigation, disposal over sloping land, or disposal through lagooning of sewage sludge.

Landfills are principally disposal sites for municipal refuse and some industrial wastes. Municipal refuse is generally composed of 40 to 50% (by weight) of organic matter, with the remaining consisting of moisture and inorganic matter such as glass, cans, plastic, pottery, etc. Under aerobic decomposition, carbonic acid that is formed reacts with any metals present and calcareous materials in the rocks and soil, thus increasing the hardness and metal content of the leachate. Decomposition of the organic matter also produces gases, including CO_2, CH_4, H_2S, H_2, NH_3 and N_2, of which CO_2 and CH_4 are the most significant soil contaminants (2).

3.0 ENVIRONMENTAL FACTORS AND CHEMISTRY

3.1 General

The chemical, physical and biological properties of a substance in conjunction with the environmental characteristics of an area, result in physical, chemical, and biological processes associated with the transport and transformation of the substance in soil and groundwater. These processes are shortly described in the following sections, along with some representative mathematical methods or models employed in the literature. Information is mainly obtained from Bonazountas and Fiksel (3).

The rates of each of the environmentally important chemical processes are influenced by numerous parameters, but most processes are described mathematically by only one or two variables. For example, the rate of biodegradation varies for each chemical with time, microbial population characteristics, temperature, pH, and other reactants. In modeling efforts, however, this rate can be approximated by a first-order rate constant (in units of time).

Soil models tend to be based on first-order kinetics; thus, they employ only first-order rate constants with no ability to correct these constants for environmental conditions in the simulated environment which differ from the experimental conditions. This limitation is both for reasons of expediency and due to a lack of the data required for alternative approaches. In evaluating and choosing appropriate unsaturated zone models, the type, flexibility, and suitability of methods used to specify needed parameters should be considered.

3.2 Physical Processes

The physical behavior of a chemical determines how the chemical partitions among the various environmental media and has a large effect on the environmental fate of a substance. For example, the release into soil of two different acids (with similar chemical behavior) may result in one chemical mainly volatilizing into the air and the other chemical becoming mainly sorbed to the organic fraction of the soil. The physical behavior of a substance therefore can have a large effect on the environmental fate of that substance.

The processes and corresponding physical parameters that are important in determining the behavior and fate of a chemical are different in analysis of trace-level contaminants than analyses of contaminants from large-scale releases (e.g., spills).

The processes of advection, diffusion, sorption and volatilization are most important to both trace-level analyses, and large-scale release analyses. Bulk properties (e.g., viscosity, solubility) are usually only important in simulations involving large amounts of contaminants.

Sorption/Ion-Cation Exchange

Adsorption is the adhesion of pollutant ions or molecules to the surface or soil solids, causing an increase in the pollutant concentrations on the soil surface over the concentration present in the soil moisture. Adsorption occurs as a result of a variety of processes with a variety of mechanisms and some processes may cause an increase of pollutant concentration within the soil solids — not merely on the soil surface. Adsorption and desorption can drastically retard the migration of pollutants in soils, therefore, knowledge of this process is of importance when dealing with contaminant transport in soil and groundwater. The type of pollutant will determine to what kinds of material the pollutant will sorb. For organic compounds, it appears that partitioning between water and the organic carbon content of soil is the most important sorption mechanism (4).

Sorption and desorption are usually modeled as one fully reversible process, although hystersis is sometimes observed. Four types of equations are commonly used to describe sorption/desorption processes: Langmuir, Freundlich, overall and ion or cation exchange. The Langmuir isotherm model was developed for single layer adsorption and is based on the assumption that maximum adsorption corresponds to a saturated monolayer of solute molecules on the adsorbent surface, that the energy of adsorption is constant, and that there is no transmigration of adsorbate on the surface phase.

The Langmuir model is described by:

$$ds/dt = K_{sw} \cdot (s_e - s)$$
$$s_e = Q° \cdot b \cdot c/(1 + c)$$

The Freundlich sorptive isotherm is an empirical model expressed by:

$$s = x/m = Kc^{1/n}$$

where: ds/dt = temporal variation of adsorbed concentration of compound on soil particles; s = adsorbed concentration of compound on soil particles; K_{sw} = Langmuir equilibrium soil-water adsorption kinetic coefficient; s_e = maximum soil adsorption capacity; $Q°$ = number of moles (or mass) of solute adsorbed per unit weight of adsorbent (soil) during maximum saturation of soil; b = adsorption partition coefficient; t = time; c = concentration of pollutant in soil moisture; x = adsorbed pollutant mass on soil; m = mass of soil; K = adsorption (partitioning) coefficient; c = dissolved concentration of pollutant in soil moisture; n = Freundlich equation parameter. At trace levels, many substances (particularly organics) are simply proportional to concentration, so the Freundlich isotherm is frequently used with $1/n = 1$. For organics, K_{oc} (the adsorption coefficient on organic carbon) is often used instead of K. These coefficients are related by: $K = K_{oc} \cdot$ (% organic carbon in the solid)/100.

Ion exchange (an important sorption mechanism for inorganics) is viewed as an exchange with some other ion that initially occupies the adsorption site on the solid. For example, for metals (M^{++}) in clay the exchanged ion is often calcium.

$$M^{++} + [clay] \cdot Ca \qquad Ca^{++} + [clay] \cdot M$$

Cation exchange can be quite sensitive to other ions present in the environment. The calculation of pollutant mass immobilized by cation exchange is given by:

$$S = EC \cdot MWT/VAL$$

where: S = maximum mass associated with solid (mass pollutant/mass of soil); EC = cation exchange capacity (mass equivalents/mass of dry soil); MWT = molecular (or atomic) weight of pollutant (mass/mole); VAL = valence of ion (−). For additional details see Bonazountas and Wagner (5).

Diffusion/Volatilization

Diffusion in solution is the process whereby ionic or molecular constituents move under the influence of their kinetic activity in the direction of their concentration gradient. The process of diffusion is often known as self-diffusion, molecular diffusion, or ionic diffusion. The mass of diffusing substance passing through a given cross section per unit time is proportional to the concentration gradient (Fick's first law).

Volatilization refers to the process of pollutant transfer from soil to air and is a form of diffusion, the movement of molecules or ions from a region of high concentration to a region of low concentration. Volatilization is an extremely important pathway for many organic chemicals, and rates for volatilization from soil vary over a large range. This process is less important for inorganic than for organic chemicals; most ionic substances are usually considered to be non-volatile (4).

Many models are available in the literature, and some of these models can be applied only to specific environmental situations and only for chemicals for which they were developed. Obviously, all models do not provide the same numerical results when employed to provide answers to a particular problem, so care must be taken in choosing an appropriate unsaturated zone model, or when specifying a volatilization rate. For modeling algorithms, and numerical examples the reader is referred to the work of Lyman et al. (6), Bonazountas & Wagner (5) and others listed in these references.

3.3 Chemical Processes

Ionization

Ionization is the process of separation or dissociation of a molecule into particles of opposite electrical charge (ions). The presence and extent of ionization has a large effect on the chemical behavior of a substance. An acid or base that is extensively ionized may have markedly different solubility, sorption, toxicity, and biological characteristics than the corresponding neutral compound. Inorganic and organic acids, bases, and salts may be ionized under environmental conditions. A weak acid HA will ionize to some extent in water according to the reaction:

$$HA + H_2O \rightleftharpoons H_3O^+ + A^-$$

The acid dissociation constant K_a is defined as the equilibrium constant for this reaction:

$$K_a = [H_3O^+] [A^-]/[HA] [H_2O]$$

Note that a compound is 50% dissociated when the pH of the water equals the pK_a ($pK_a = - \log Ka$).

Hydrolysis

Hydrolysis is one of a family of reactions which leads to the transformation of pollutants. Under environmental conditions, hydrolysis occurs mainly with organic compounds. Hydrolysis is a chemical transformation process in which an organic RX reacts with water, forming a new molecule. This process normally involves the formation of a new carbon-oxygen bond and the clearing of the carbon-X bond in the original molecule:

$$RX \xrightarrow{\text{H}_2\text{O}} R-OH + X^- + H^+$$

Hydrolysis reactions are usually modeled as first-order processes, using rate constants (K_H) in units of (time) $^{-1}$:

$$-d[RX]/dt = K_H[RX]$$

The rate of hydrolysis of various organic chemicals, under environmental conditions can range over 14 orders of magnitude, with associated half-lifes (time for one-half of the material to disappear) as low as a few seconds to as high as 10^6 years and is pH dependent. It should be emphasized that if laboratory rate constant data are used in soil models and not corrected for environmental conditions — as is often the only choice — then model results should be evaluated with skepticism.

Oxidation/Reduction

For some organic compounds, such as phenols, aromatic amines, electron-rich olefins and dienes, alkyl sulfides, and eneamines, chemical oxidation is an important degradation process under environmental conditions. Most of these reactions depend on reactions with free-radicals already in solution and are usually modeled by pseudo-first-order kinetics:

$$-d[X]/dt = K'_O [RO_2\cdot] [X] = K_{OX} [X]$$

where: X is the pollutant, K'_O is the second order oxidation rate constant, $RO_2\cdot$ is a free radical, and K_{OX} is the pseudo-first-order oxidation rate constant.

Complexation

Complexation, or chelation, is the process by which metal ions and organic or other non-metallic molecules (called ligands) can combine to form stable metal-ligand complexes. The complex that is found will generally prevent the metal from undergoing other reactions or interactions that the free metal cation would. Complexation may be important in some situations; however, the current level of understanding of the process is not very advanced, and the available information has not been shown to be particularly useful to quantitative modeling (5).

3.4 Biological Processes

Bioaccumulation is the process by which terrestrial organisms such as plants and soil invertebrates accumulate and concentrate pollutants from the soil. Bioaccumulation is not examined in soil modeling, aside from some nutrient cycle (phosphorus, nitrogen) and carbon cycle bioaccumulation attempts.

Biodegradation refers to the process of transformation of a chemical by biological agents, usually by microorganisms and it actually refers to the net result of a number of different processes such as: mineralization, detoxication, cometabolism, activation, and change in spectrum. In toxic chemical modeling, biodegradation is usually treated as a first-order degradation process ($\underline{5}$):

$$dc/dt = -K_{DE} \cdot c^n$$

where: c = dissolved concentration of pollutant soil moisture (ug/mL); K_{DE} = rate of degradation (day^{-1}); and n = order of the reaction ($n = 1$; first order).

4.0 MATHEMATICAL MODELING

4.1 General Overview

Pollutant fate mathematical modeling in soil systems is an area of current intensive work, because of the numerous problems originating at hazardous waste sites. The variety of models has dramatically increased during the last decade, but although the variety of models appears to be large only very few "different" modeling concepts exist and very few physical or chemical processes are modeled. In general, soil/groundwater modeling concepts deal mainly with point source pollution and can be categorized into: (1) unsaturated soil zone (or soil); (2) saturated soil zone (groundwater); (3) geochemical, and (4) ranking. The first two categories follow comparable patterns of mathematics and approach, the third enters into chemistry and speciation modeling, and the fourth follows a screening approach as discussed in section 4.4.

There is no scientific reason for a soil model to be an unsaturated soil model only, and not to be an unsaturated (soil) *and* a saturated soil (groundwater) model. Only mathematical complexity mandates the differentiation, because such a model would have to be 3-dimensional (e.g., $\underline{7}$) and very difficult to operate. Most of the soil models account for vertical flows, groundwater models for horizontal flows.

In summary, models can be classified in general into: deterministic, which describe the system as cause/effect relationships; and stochastic, which incorporate the concept of risk, probability or other measures of uncertainty. Deterministic and stochastic models may be developed from: observation, semi-empirical approaches, and theoretical approaches. In developing a model, scientists attempt to reach an optimal compromise among the above approaches, given the level of detail justified by both the data availability and the study objectives. Deterministic model formulations can be further classified into: simulation models which employ a well accepted empirical equation, that is forced via calibration coefficients, to describe a system; and analytic models in which the derived equation describes the physics/chemistry of a system.

Without a solution, formulated mathematical systems (models) are of little value. Four solution procedures are mainly followed: the analytical, the numerical (e.g., finite different, finite element), the statistical, and the iterative. Numerical techniques have been standard practice in soil quality modeling. Analytical techniques are usually employed for simplified and idealized situations. Statistical techniques have academic respect, and iterative solutions are developed for specialized cases. Both the simulation and the analytic models can employ numerical solution procedures for their equations. Although the above terminology is not standard in the literature, it has been used here as a means of outlining some of the concepts of modeling.

Generally speaking, a deterministic or stochastic soil quality model consists of two major parts of modules:

    (1) The flow module or moisture module, or hydrologic cycle module — aiming to predict flow or moisture behavior (i.e., velocity, content) in the soil; and

    (2) The solute module — aiming to predict pollutant transport, transformation and soil quality in the soil zone.

The above two modules form the soil quality model. The flow module drives the solute module. It is important to note that the moisture module can be absent from the model and in this case a model user has to input to the solute module information that would have been either produced by a moisture module, or would have been obtained from observed data at a site.

At this stage of intensive research in soil and groundwater quality modeling it may be reasonable to group the prevailing modeling concepts of the literature into three major categories: the "Traditional Differential Equation (TDE) modeling," the "compartmental" modeling, and "other" types. This terminology is again not a standard practice, but is employed here for reasons of communication. TDE modeling applies to both the flow or moisture module and the solute module, and a modeling package may consist of one TDE module (e.g., moisture) and a compartmental solute module, or vice versa.

The coming sections aim to clarify some key issues related to soil and groundwater models. The following documents provide an overview of this area of science: The series of articles by Mercer & Faust (8) describing groundwater modeling concepts are equally applicable to unsaturated soil zone modeling; the monograph of Bachmat et al. (9) listing various models; the work of Bonazountas & Wagner (5) introducing the compartmental soil quality modeling concept and geochemical modeling; the reference book of Freeze & Cherry (10); and the modeling handbook and catalogue in preparation by Bonazountas & Fiksel (3). Reference to the above sources is not meant to exclude other excellent publications presented in reputable scientific journals; rather it indicates selected basic sources employed to draft the coming sections.

In the following sections more emphasis is placed on the unsaturated soil zone than on groundwater modeling. This emphasis can be justified by the fact that similar modeling concepts govern both environments.

4.2 Unsaturated Soil Zone (Soil) Modeling

4.2.1 TDE Modeling

Soil modeling follows three different mathematical formulation patterns: (1) Traditional Differential Equation (TDE) modeling; (2) Compartmental modeling; and (3) Stochastic modeling. Some researchers may categorize models differently as for example into numerical or analytic, but this categorization applies more to the techniques employed to solve the formulated model, rather than to the formulation per se.

The TDE moisture module (of the model) is formulated from three equations: (1) the water mass balance equation, (2) the water momentum, (3) the Darcy equation, and (4) other equations such as the surface tension of potential energy equation. The resulting differential equation system describes moisture movement in the soil and is written in a one dimensional, vertical, unsteady, isotropic formulation as:

$$\partial[K(\psi) \, (\partial\psi/\partial z \, + \, 1)]/\partial z \, = \, C(\psi)\partial\psi/\partial t \, + \, S \tag{1}$$

$$v_z \, = \, - \, K \, (z,\psi)\partial\phi/\partial z \tag{2}$$

where: z = elevation (cm); ψ = pressure head, often called soil moisture tension head in the unsaturated zone (cm); K(ψ) = hydraulic conductivity (cm/min); C(ψ) = dθ/dψ = slope of the moisture (θ) versus pressure head (ψ) (cm^{-1}); t = time (min); S = water source or sink term (min^{-1}); φ = z + ψ; and v$_z$ = vertical moisture flow velocity (cm/sec). The moisture module output provides the parameters v and θ as input to the solute module.

The TDE solute module is formulated with one equation describing pollutant mass balance of the species in a representative soil volume dV = dxdydz. The solute module is frequently known as the dispersive, convective differential mass transport equation, in porous media, because of the wide employment of this equation, that may also contain an adsorptive, a decay and a source or sink term. The one dimensional formulation of the module is:

$$\partial(\theta c)/\partial t = [\partial(\theta \cdot K_D \partial c/\partial z)/\partial t] - [\partial(vc)/\partial z] - [\rho \cdot \partial s/\partial t] \pm \Sigma P \tag{3}$$

where: θ = soil moisture content; c = dissolved pollutant concentration: in soil moisture; K$_D$ = apparent diffusion coefficient of compound in soil-air; v = Darcy velocity of soil moisture; ρ = soil density; s = adsorbed concentration of compound on soil particles; ΣP = sum of sources or sinks of the pollutant within the soil volume; and z = depth.

Models of the above have been presented by various researchers of the U.S. Geological Survey (USGS) and the academia. The above equation has been solved principally: (a) numerically over a temporal and spatial discretized domain, via finite difference or finite element mathematical techniques (e.g., 11); (b) analytically, by seeking exact solutions for simplified environmental conditions (e.g., 12); or (c) probabilistically (e.g., 13).

At this point it is important to note that the flow model (a hydrologic cycle model) can be absent from the overall model. In this case the user has to input to the solute module [i.e., equation (1)] the temporal (t) and spatial (x,y,z) resolution of both the flow (i.e., soil moisture) velocity (v) and the soil moisture content (θ) of the soil matrix. This approach is employed by Enfield et al. (12) and other researchers. If the flow (moisture) module is not absent from the model formulation (e.g., 14), then the users are concerned with input parameters, that may be frequently difficult to obtain. The approach to be undertaken depends on site specificity and available monitoring data.

Some principal modeling-specific deficiencies when modeling solute transport via the TDE approach are: (1) only diffusion, convection, adsorption and possibly decay can be modeled, whereas processes such as fixation or cation-exchange have to be either neglected or represented with the sources and sinks term of the equation because of mathematical complexity; (2) this equation is applicable mainly to pollutant transport of organics, whereas transport of metals which can be strongly affected by other processes cannot be directly modeled; (3) this equation can predict volatilization only implicitly via boundary diffusion constraints; however, experimental studies have frequently demonstrated an over-estimation or underestimation of the theoretical volatilization rate unless a "sink" or source term is included in the equation; (4) no experimental or well accepted equation for a process (e.g., volatilization) can be incorporated since the model has its own predictive mechanism; (5) pollutant concentrations are estimated only in the soil-moisture and on soil-particles, whereas pollutant concentrations in the soil-air are omitted; and (6) the discretized version of the equation in case of numerical solutions has a pre-set temporal and spatial discretization grid, that results in high operational costs (professional time, computer time) of the model, since input data have to be entered to the model for each node of the grid.

In a recent modeling evaluation effort, Murarka (15) reports that the currently available coupled or uncoupled models of hydrologic flow and the geochemical interactions are adversely affected by the following factors: difficulties in establishing consistency between the theoretical frameworks, laboratory experiments, and field research; limited basic knowledge about nonequilibrium conditions and phase relations; inadequate existence of geochemical submodels to couple with the hydrologic transport submodels; uncertainties in input data particularly for dispersion and chemical reactions rate coefficients; and numerical difficulties with model solution techniques.

4.2.2 Compartmental Modeling

Compartmental soil modeling is a new concept and can apply to both modules. For the solute fate module, for example, it consists of the application of the law of pollutant mass conservation to a representative user specified soil element. The mass conservation principle is applied over a specific time step, either to the entire soil matrix or to the subelements of the matrix such as the soil-solids, the soil-moisture and the soil-air. These phases can be assumed in equilibrium at all times; thus once the concentration in one phase is known, the concentration in the other phases can be calculated. Single or multiple soil compartments can be considered whereas phases and subcompartments can be interrelated (Figure 2) with transport, transformation and interactive equations.

Compartmental models may by-pass the deficiencies of the TDE modeling because they may handle geochemical issues in a more sophisticated way if required, but this does not imply that compartmental models are "better" than TDE models. They are simply different. Compartmental models reflect the personal "touch" of their developers and cannot be formulated under generalized guidelines or concepts. The moisture module (i.e., driving element) of a compartmental solute model can be either incorporated into the overall model, or can be an independent module, as for example a TDE module of the literature. At this stage of scientific research, the most developed soil compartment model appears to be SESOIL: Seasonal Soil Compartment model (5). SESOIL consists of a dynamic compartment moisture module, and a dynamic compartmental solute transport module. The following paragraphs present a demonstration of the basic mathematical equations governing compartmental soil quality modeling. This information has been abstracted from the SESOIL model.

The law of pollutant mass concentration for a representative element can be written over a small time step as:

$$\Delta M = M_{in} - M_{out} - M_{trans} \qquad (4)$$

The solute (dissolved) concentration of a compound can be related to its soil-air concentration via Henry's law:

$$c_{sa} = c \cdot H/R \; (T + 273) \qquad (5)$$

where: c_{sa} = pollutant concentration in soil-air; c = dissolved pollutant concentration; H = Henry's law constant; R = gas constant; T = temperature in °C; and °K = °C + 273.

The pollutant concentration of the *soil* (i.e., solids) can be determined from the *sum* of the concentrations of the pollutant adsorbed, cation-exchanged, and/or other-

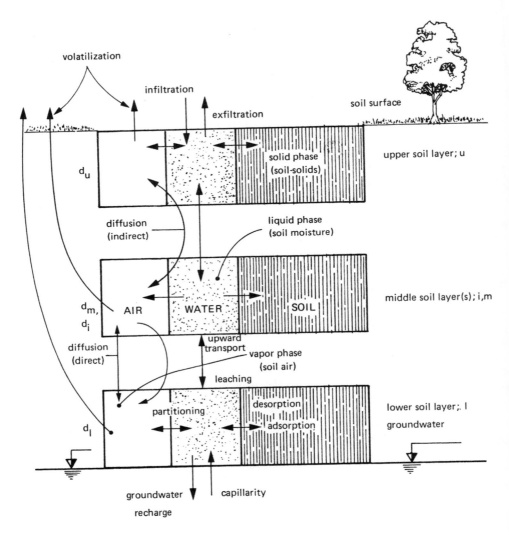

Figure 2.   Schematic of phases in soil matrix. (Reproduced
with permission from Ref. 5.)

wise associated with the soil particles, e.g., via adsorption isotherms. One commonly used adsorption isotherm equation is the Freundlich equation:

$$s = K \cdot c^{1/n} \tag{6}$$

where: s = adsorbed concentration of compound; K = partitioning coefficient; c = dissolved concentration of compound; and n = Freundlich constant.

The total concentration of a chemical in a soil matrix can be calculated from the concentration of the pollutant in each phase and the related volume of each phase by:

$$c_o = (n - \theta)\, c_{sa} + (\theta)\, c + (\rho_b)\, s \tag{7}$$

where: c_o = overall (total) concentration of pollutant in soil matrix; n = soil (total) porosity; θ = soil moisture content; $(n - \theta) = n_{air}$ soil-air content or soil-air filled porosity; c_{sa} = pollutant concentration in soil-air; c = pollutant concentration (dissolved) in soil-moisture; ρ_b = soil bulk density; and s = pollutant concentration on soil particles.

The above expressions are input terms to equation (4), which is then applied for each time step, each subcompartment, and each compartment of the user specified matrix (Figure 2). The term M_{in} may reflect input pollution from rain (upper layer), from soil-moisture from an above layer, and from a below layer. The term M_{out} reflects pollution exports from the individual compartment, whereas the term M_{trans} reflects all transformation and chemical reactions taking place in the compartment. All terms can be normalized to be soil moisture concentration via interconnecting equations such as (5), and (6) which can describe processes such as volatilization, cation exchange, photolysis, degradation, hydrolysis, fixation, biologic activity, etc. The solution of the resulting system of equations is a complicated issue, and may require — for computational efficiency and other reasons — development of new numerical solution techniques or algorithms (e.g., SESOIL model).

4.2.3 Stochastic, Probabilistic, Other Modeling Concepts

Stochastic or probabilistic techniques can be applied to either the moisture module, or the solution of equation (3) — or for example the models of Schwartz & Crowe (13) and Tang et al. (16), or can lead to new conceptual model developments as for example the work of Jury (17). Stochastic or probabilistic modeling is mainly aimed at describing "breakthrough" times of overall concentration threshold levels, rather than individual processes or concentrations in individual soil compartments. Coefficients or response functions and these models have to be calibrated to field data since major processes are studied via a black-box or response function approach and not individually. Other modeling concepts may be related to soil models for solid waste sites and specialized pollutant leachate issues (18).

4.2.4 Physical, Chemical, Biological Processes Modeled

Modelers should be fully aware of the range of applicability and processes considered by a computerized package. There exists some disagreement among soil modelers as to whether there is a need for increased model sophistication, since almost all soil modeling predictions have to be validated with monitoring data, given the physical, chemical and biological processes that affect pollutant fate in soil systems. Because of the latter consideration, many simplified models may provide excellent results, assuming accurate site specific calibration is achieved. Nevertheless model sophistication is reflected in the processes modeled, but model selection is mandated by the project needs and data availability.

The important physical processes of a typical soil model are:

(1) The hydrologic cycle, or moisture cycle — that may encompass the processes of rain infiltration in the soil, exfiltration from the soil to the air, surface runoff, evaporation, moisture behavior, groundwater recharge and capillary rise from the groundwater. All these processes are interconnected and are frequently referred to as the hydrologic cycle components.

(2) The pollutant or solute cycle — that may encompass the processes of advection, diffusion, volatilization, adsorption and desorption, chemical degradation or decay, hydrolysis, photolysis, oxidation, cation or anion exchange, complexation, chemical equilibria, nutrient cycles, and others (see section 3.0).

(3) The biological cycle — that may encompass processes of biological transformation, plant uptake, bioaccumulation, soil organisms transformations and others.

Models of the literature can handle one or more of the above processes and for various pollutants. In general, however, soil models tend to handle:

(1) From the hydrologic cycle: temporal resolution of soil moisture surface, runoff, and groundwater recharge components, by inputting to the model the "net" infiltration rate into the soil column; and

(2) From the pollutant and biological cycles: the processes of advection, diffusion, volatilization (diffusion at the soil-air interface), adsorption or desorption (equilibrium), and degradation or decay, which are also the most important chemical processes in the soil zone. All other processes can be "lumped" together under the source or sink term of equation (3).

Fortunately — and not unfortunately — no one model exists as yet which simulates all of the physical, chemical, and biological processes associated with pollutant fate in soils. We say fortunately, because such a package would be very data intensive and difficult to use. Intensive research is required to accomplish the above objective and the value of the overall product may be questioned by users. Section 7.0 presents selected models.

4.3 Saturated Soil Zone (Groundwater) Modeling

Saturated soil zone (or groundwater) modeling is formulated almost exclusively via a TDE system, consisting of two modules, the flow and the solute module. The two modules are written as (9):

$$\nabla \cdot \frac{\rho k}{\mu}(\nabla p - \rho g \nabla Z) - q = \frac{\partial}{\partial t}(\phi \rho) \tag{8}$$

$$\nabla \cdot (\rho C \frac{k}{\mu}(\nabla p - \rho g \nabla Z)) + \nabla \cdot (\rho E) \cdot \nabla C - qC = \frac{\partial}{\partial t}(\rho \phi C) \tag{9}$$

where: C = concentration, mass fraction; E = dispersion coefficient; g = acceleration due to gravity; k = permeability; p = pressure; q = mass rate of production or injection of liquid per unit volume; t = time; Z = elevation above a reference plane; ϕ = porosity; ρ = density; and μ = viscosity.

Mathematical groundwater modeling has been the least problematic in its scientific formulation, but has been the most problematic model category when dealing with applications, since these models have to be calibrated and validated as described later. Actually we have only TDE and some other (e.g., stochastic) formulations. The proliferation of literature models (9) is mainly due to the: different model dimensionalities (zero, one, two, three); model features (e.g., with adsorption, without absorption terms); solution procedures employed (e.g., analytic, finite difference, finite element, random walk, stochastic) for equation system (7) and (8); sources and sinks described; and the variability of the boundary conditions imposed. Some of the principal modeling deficiencies discussed in the previous section (soil models) apply to groundwater models also. In general: (1) there exists no "best" groundwater model, and (2) for site specific applications groundwater models have to be calibrated.

The two principal solution methods for equations (8) and (9) that result in different model categories with substantially different impacts on the level of effort required to run a package are the: (1) analytical models; and (2) numerical models. Employment of the first method results in an analytic expression applicable to the entire groundwater domain. Employment of the second method results in formulation of expressions applicable to the nodes or the elements of a domain, the number of nodes or elements being user specified. In analytical modeling only averaged data for the entire domain have to be input to the model. The numerical modeling data have to be input for all nodes or elements of the model, a fact that frequently results in high model cost runs, in terms of both professional and computational time. Common numerical solution techniques are the finite difference, the finite element, the method of characteristics, the random walk, and their variations. Interested readers are referred to Mercer & Faust (8), Prickett et al. (19), and Bear (20).

4.4 Ranking Modeling

Ranking models are aimed at assessing environmental impacts of waste disposal sites. The first ranking models focused on groundwater contamination; later models had a wider scope (e.g., health considerations). These models rank or rate contaminant migration at different sites, as it is affected by hydrogeologic, soil, waste type, density, and site design parameters. These models are based on questions and answers and on weighting factors the user has to specify. They are very subjective in their use, and their output is frequently difficult to justify scientifically. Despite the above facts, they have received a wide dissemination because they are easy to use and do not require use of computers. Well known models are the: LeGrand (21), Silka & Swearingen (22), JRB (23), MITRE (24), and Arthur D. Little, Inc. (25). Interested readers should refer to the original publications.

4.5 Aquatic Equilibria Models

An evaluation of the fate of trace metals in surface and sub-surface waters requires more detailed consideration of complexation, adsorption, coagulation, oxidation-reduction, and biological interactions. These processes can affect metals, solubility, toxicity, availability, physical transport, and corrosion potential. As a result of a need to describe the complex interactions involved in these situations, various models have been developed to address a number of specific situations. These are called equilibrium or speciation models because the user is provided (model output) with the distribution of various species.

There are two basic approaches to the solution of a species distribution problem: (1) The equilibrium constant approach, and (2) the Gibbs free energy approach. Most models use the former approach, which utilizes measured equilibrium constants for all mass action expressions of the systems. The latter approach uses free energy values. In both cases, the most stable condition is sought, and a solution to a set of nonlinear equations is required. The solution involves an iterative procedure, as discussed by Nordstrom et al. (26). The use of the Gibbs free energy minimization approach is primarily useful for simple systems due to the limited availability of free energy values. The larger data base for equilibrium constants makes this method generally preferable. GEOCHEM (27) is an aquatic geochemical equilibrium model. Aquatic equilibrium models are at a developmental stage. Current versions are steady-state models, and are formulated for one soil compartment.

5.0 SELECTED MODELS AND ISSUES

Table 3 lists selected soil and groundwater models and their main features. Table 4 lists limitations and advantages of major model categories. Models listed in Table 3 are documented, operational and very representative of the various structures, features and capabilities. For example:

(1) PESTAN (12) is a dynamic TDE soil solute (only) model, requiring the steady-state moisture behavior components as user input. The model is based on the analytic solution of equation (3), and is very easy to use, but has also a limited applicability, unless model coefficients (e.g., adsorption rate) can be well estimated from monitoring studies. Moisture module requirements can be obtained by any model of the literature.

(2) SCRAM (28) is a TDE dynamic, numerical finite difference soil model, with a TDE flow module and a TDE solute module. It can handle moisture behavior, surface runoff, organic pollutant advection, dispersion, adsorption, and is designed to handle (i.e., no computer code has been developed) volatilization and degradation. This model may not have received great attention by users because of the large number of input data required.

(3) SESOIL (5,29) is a dynamic soil compartmental model; with a hydrologic cycle and a pollutant cycle compartmental structure, that permits users to tailor the model temporal and spatial resolution to the study objectives. The model estimates the hydrologic cycle components (including moisture behavior) from available NOAA, USDA, and USGS data, and simulates the pollutant cycle by accounting for a number of chemical processes for both inorganic (metal) and organic pollutants.

(4) PATHS (30) is mainly an analytical groundwater model, that provides a rough evaluation of the spatial and temporal status of a pollutant fate. The model has its own structure and features, and is a deviation from the TDE, or the compartmental, or stochastic approaches.

(5) AT123D (31) is a series of soil or groundwater analytical submodels, each submodel addressing pollutant transport: in 1-, 2-, or 3-dimensions; for saturated or unsaturated soils; for chemical, radioactive waste heat pollutants; and for different types of releases. The model can provide up to 450 submodel combinations in order to accommodate various conditions analytically.

TABLE 3
SELECTED MODELS AND FEATURES[1]

Continued on next page

Model Acronym	Model Type[2]				Model Formulation[3]					Mathematics[4]			Chemistry Issues[5]					User Concerns[6]				Contact/Information
	Unsaturated Zone	Groundwater	Aquatic Equilibrium	Ranking	Flow Module	Solute Module	TDE Approach	Compartmental	Statistical, Other	Analytical	Numerical	Statistical	Organics	Inorganics	Metals	Gaseous Phase	Increased Chemistry	Input Data Requirements	Calibration	Level of Effort	Application Study	
PESTAN	•					•	•			•			•					L	L	L	•	Enfield; EPA (405) 332-8800
SCRAM	•					•	•				•		•					H	H	H	•	EPA Report PB-259933 (Adams & Kuzisu, 1976)
SESOIL	•				•	•		•	•		•		•	•	•	•	•	M	M	M	•	Bonazountas; ADL (617) 864-5770
AT123D	•	•			•	•				•			•	•				M	L	M	•	Yeh; ORNL (615) 574-7285
PLUME		•			•	•					•		•	•				L	L	L	•	Wagner; OSU (405) 624-5280
PATHS		•			•	•				•	•		•	•				L	M	M	•	Nelson; Battelle (509) 376-8332
MMT/VVT		•			•	•					•		•	•				H	M	M	•	Cole; Battelle (509) 376-8451
FEM WASTE		•			•	•					•		•					H	H	H	•	Yeh; ORNL (615) 574-7285
R. WALK		•				•	•			•			•					H	H	H		Prickett; ILIOS (217) 344-2277
USGS Models		•			•	•	•				•		•	•	•	•		H	H	H	•	Appel; USGS (703) 860-6892
GEO CHEM			•					•						•	•			H	L	L		Mattigod; UCR (714) 787-1012
MITRE/JRB	•	•		•										•					L	M	•	MITRE (703) 827-6000; JRB (703) 821-4873
ADL/LeGrand	•	•		•														M	M	M	•	ADL (617) 864-5770; (919) 787-5855

Table 3 Notes:

1. This is a partial list of available well documented models. An EPA Modeling Handbook and Catalogue under preparation (Bonazountas & Fiksel, 1982) will list additional models.

2. The use of complex (e.g., a numerical soil and groundwater package) models that can handle more than one compartment is not always desirable, since generalized packages tend to be cumbersome, unless especially designed (class of models).

3. The most representative characteristics are given. The Traditional Differential Equation (TDE) approach applies to the flow and solute module. Under "other" we may have for example linear analytic system solutions.

4. The most representative characteristics are given, since for example statistical formulations can be subject to statistical analytic or numerical solution procedures.

5. Almost all models can simulate organic, inorganic and metal fate, assuming that a careful calibration via an adsorption coefficient may alter the model output to predict measured/monitored values. However, not all models have by design increased chemistry capabilities (e.g., cation exchange capacity; complexation), therefore, the most representative capabilities are indicated.

6. L = low, M = medium, H = high input data requirements. In general, numerical models have higher input data requirements and calibration needs, therefore, may better represent spatial resolution of a domain. Compartment models provide an optimal compromise (see Section 4.2). The level of effort is intuitively defined here.

TABLE 4

MOST IMPORTANT CHARACTERISTICS OF MAJOR MODEL CATEGORIES

Model Category	Advantages	Disadvantages	Comments
Soil and Groundwater Models			
* TDE – Type	Clear Formulation/Capabilities	Rigid Model Structure, Limited Capabilities	This has been the traditional computerized modeling approach
– Analytic	Easy model use; limited calibration possibilities; limited input data requirements; desk computer use	Rough averaged predictions of pollutant fate, limited application capabilities	To be used as an overall fate (screening) tools
– Numerical	Wide range of applications; detailed spatial, temporal resolutions; increased chemistry capabilities	Extensive calibration requirements; input data intensive (nodes, elements, time); require computer use and related skills	Recommended for site specific applications
* Compartmental Type	Can be tailored to user's requirements; increased chemistry capabilities; can better meet spatial and temporal domain requirements	Expected user interaction and problem understanding	Today's scientific tendency
Aquatic Equilibrium Model	Increased chemistry capability	Data intensive, parameters may not be available	Models at a developmental stage
Ranking Models	Easy to use with available data	Simplistic approach, output reflects user's intuition	Employed by the EPA, U.S. Army, Air Force and Navy

(6) MMT (32) is a 1- or 2-dimensional solute transport numerical groundwater model, to be driven off-line by a flow transport, such as VTT (Variable Thickness Transport). MMT employs the random-walk numerical method and was originally developed for radionuclide transport. The model accounts for advection, sorption and decay.

The remaining models of Table 1 follow the scientific basic patterns described above with small variations. All models handle one species at a time, and two soil models (SESOIL, AT123D) can handle gaseous pollutants also. Other documented soil models of the literature are the package (TDE approach) of Perez et al. (33), the work (TDE approach) of Feddes et al. (34), and the models of Freeze (10), Van Genuchten & Pinder (35), and Narasimhan (7). The U.S. Geological Survey (USGS) has been very active for a number of years in TDE soil and groundwater quality model development (36). Models of the USGS are well documented and available in the public domain.

Model selection, application and validation are issues of major concern in mathematical soil and groundwater quality modeling. For the model selection, issues of importance are: the features (physics, chemistry) of the model; its temporal (steady state, dynamic) and spatial (e.g., compartmental approach resolution); the model input data requirements; the mathematical techniques employed (finite difference, analytic); monitoring data availability; and cost (professional time, computer time). For the model application, issues of importance are: the availability of realistic input data (e.g., field hydraulic conductivity, adsorption coefficient); and the existence of monitoring data to verify model predictions. Some of these issues are briefly discussed below.

Input data have to be compiled and input to the model from: site specific investigations and analyses (e.g., leaching rates of pollutants, soil permeability); national data bases (e.g., climatological data from the NOAA; and other sources (e.g., diffusion rates of pollutants from handbooks). Compilation of input data for site specific computer runs are model specific, geohydrology and chemistry specific. Some data categories are: pollutant source data, climatological data, geographic data, particulate transport data and biological data.

Exact knowledge of the physics of the soil system — although essential — is impossible prior to employing any modeling package. Numerical (e.g., finite difference) TDE soil models, for example, require the net infiltration rainfall rate after each storm event as an input parameter to their moisture module. This rate can be either a user input or can be generated by another model. The same models require the soil conductivity as a function of the soil moisture content as an input parameter. Its value can be obtained either from field investigations, or from laboratory data, or from references, but much uncertainty exists in this area of input data gathering (37).

Numerical soil models (time, space) provide a general tool for quantitative and qualitative analyses of soil quality, but require time consuming applications that may result in high study costs. In addition input data have to be given for each node or element of the model, which model has to be run twice, the number of rainfall events. On the other hand, analytic models obtained from analytic solutions of equation (3) are easier to use, but can simulate only averaged temporal and spatial conditions, which may not always reflect real world situations. Statistical models may provide a compromise between the above two situations.

Model output "validation" is essential to any soil modeling effort, although this term has a broad meaning in the literature. For the purpose of this section we can define validation as "the process which analyzes the validity of final model output," namely the validity of the predicted pollutant concentrations or mass in the soil column (or in groundwater), to groundwater and to the air, as compared to available knowledge of measured pollutant concentrations from monitoring data (field sampling). A disagreement of course in absolute levels of concentration (predicted versus measured) does not necessarily indicate that either method of obtaining data (modeling, field sampling) is incorrect or that either data set needs revision. Field sampling approaches and modeling approaches rely on two different perspectives of the same situation.

Important issues in groundwater model validation are: the estimation of the aquifer physical properties, the estimation of the pollutant diffusion and decay coefficient. The aquifer properties are obtained via flow model calibration (i.e., parameter estimation; see Bear, 20), and by employing various mathematical techniques such as kriging. The other parameters are obtained by comparing model output (i.e., predicted concentrations) to field measurements: a quite difficult task, because clear contaminant plume shapes do not always exist in real life.

Three major input data categories are required for soil and groundwater modeling efforts: climatologic or hydrologic data, soil data and chemistry data. These data are used as input to models and to validate models output. Climatologic data can be obtained from site specific investigations or from NOAA or USGS records. Soil data can be obtained from site specific investigation (e.g., soil hydraulic conductivity) or from USDA data information documents. Chemistry data can be obtained from reference books (e.g., Lyman et al., 6) or from laboratory analyses (e.g., adsorption coefficient). Data are model-specific, and environment-specific.

6.0 ACKNOWLEDGMENTS

Information contained in this paper has been compiled under support provided by Melanie Byrne, Joanne Perwak, Kate Scow and Janet Wagner. Their contributions are appreciated. In addition, this paper will serve as a background section for an Environmental Modeling Catalogue (Bonazountas & Fiksel, editors, 1983) in preparation for the EPA/TIP, Mike Alford Project Officer, EPA Contract No. 68-01-5146.

LITERATURE CITED

1. Knisel, W.G., editor (1980). CREAMS. A field scale model for Chemicals, Runoff, and Erosion from Agricultural Management Systems. U.S. Department of Agriculture, Washington, DC.
2. Zanoni et al. (1973), as reported in Bonazountas and Fiksel (1982).
3. Bonazountas, M.; J. Fiksel (1982). ENVIRO: Environmental Mathematical Pollutant Fate Modeling Handbook/Catalogue, EPA Contract No. 68-01-5146, Arthur D. Little, Inc., Cambridge, MA 02140.
4. Mackay, D. et al (1982) in Convay (1982).
5. Bonazountas, M.; J. Wagner (1981). SESOIL: A seasonal soil compartment model. Office of Toxic Substances, U.S. Environmental Protection Agency, Washington, DC.
6. Lyman, W. et al. (1982). Chemical property estimation methods. McGraw-Hill Book Company, New York, NY.
7. Narasimhan, T.N. (1975). A unified numerical model for saturated-unsaturated groundwater flow. Thesis. University of California, Berkeley, Berkeley, CA.
8. Mercer, J.W.; C.R. Faust (1981). Groundwater modeling. National Water Well Association, Washington, DC.

9. Bachmat, Y., J. Bredehoeft, B. Andrews, D. Holtz and S. Sebastian (1980). Groundwater management: the use of numerical models. Water Resources Monograph 5. American Geophysical Union, Washington, DC. 127 pp.

10. Freeze, R.A.; Cherry, J.A. (1979). Groundwater. Prentice-Hall, Englewood Cliffs, NJ.

11. Mackay, D. (1979). Finding fugacity feasible. Environmental Science and Technology, Vol. 13, pp. 1218-23.

12. Enfield, C.G., R.F. Carsel, S.Z. Cohen, T. Phan and D.M. Walters (1980). Approximating pollutant transport to groundwater. USEPA. RSKERL, Ada, OK. (unpublished paper).

13. Schwartz, F.W., A. Growe (1980). A deterministic probabilistic model for contaminant transport, US:NRC, NUREG/CR-1609, Washington, DC.

14. Huff, D.D. et al. (1977). TEHM: A terrestrial ecosystem hydrology model. Oak Ridge National Laboratory, Oak Ridge, TN.

15. Murarka, L. (1982). Planning workshop on solute migration from utility solid waste, publication EA-2415, Electric Power Research Institute, Palo Alto, CA 94304.

16. Tang, D.H., F.W. Schwartz and L. Smith (1982). Stochastic modeling of mass transport in a random velocity field. Water Resources Research 18(2), pp. 231-244.

17. Jury, W.A. (1982). Simulation of solute transport using a transfer function model. Water Resources Research 18(2), pp. 363-368.

18. Schultz, D. (1982). Land disposal of hazardous waste. Proceedings of the 8th Annual Research Symposium, F.J. Hutchell, Kentucky, March 8-10. Municipal Environmental Research Laboratory, U.S. Environmental Protection Agency, Cincinnati, OH.

19. Prickett, T.A., T.G. Naymik and C.G. Lonnquist (19810. A "random walk" solute transport model for selected groundwater quality evaluations. Illinois State Water Survey, Bulletin 65.

20. Bear, J. (1979). Hydraulics of groundwater. McGraw-Hill Book Company, New York, NY.

21. Legrand, H.E. (1980). A standardized system for evaluating waste disposal sites. National Water Well Association, Washington, DC.

22. Silka, L.R. and T.L. Swearingen (1978). A manual for evaluating contamination potential of surface impoundments. U.S. EPA, Groundwater Protection Branch. EPA 570/9-78-003. 73 pp.

23. JRB Associates (1980). Methodology for rating the hazard potential of waste disposal sites. JRB Associates, McLean, VA.

24. MITRE (1981). Site ranking model for determining remedial action priorities among uncontrolled hazardous substances facilities. The MITRE Co., McLean, VA 22102.

25. Arthur D. Little, Inc. (1981). Prepared Revisions to MITRE Model, Arthur D. Little, Inc., Cambridge, MA 02140.

26. Nordstrom et al. (1979) as cited by Bonazountas and Fiksel (1982).

27. Sposito, G. (1981). The thermodynamics of soil solutions. Clarendon Press, Oxford, England.

28. Adams, R.T.; Kurisu, F.M. (1976). Simulation of pesticide movement on small agricultural watersheds. Environmental Research Laboratory, U.S. Environmental Protection Agency; Athens, GA.

29. Bonazountas, M. et al. (1981). Evaluation of seasonal soil/groundwater pollutant pathways via SESOIL: Office of Water Regulations and Standards, U.S. Environmental Protection Agency, Washington, DC.

30. Nelson, R.W.; J.A. Schur. (1980). Assessment of effectiveness of geologic oscillation systems: PATHS groundwater hydrologic model. Battelle, Pacific Northwest Laboratory, Richland, WA.

31. Yeh, G.T.; D.S. Ward (1981). FEMWASTE: A finite-element model of waste transport through saturated-unsaturated porous media. Oak Ridge National Laboratory, Environmental Sciences Division. Publication No. 1462, ORNL-5601. 137 pp.

32. Foote, H.P., (1982). For information: Battelle Pacific Northwest Laboratories, P.O. Box 999, Richland, VA 99352.

33. Perez, A.I. et al. (1974). A water quality model for a conjunctive surface, groundwater system. Office of Water Research and Technology. Environmental Protection Agency, Washington, DC.

34. Feddes, R.A. et al. (1978). Simulation of field water use. John Wiley and Sons, New York, NY.

35. Van Genuchten, M.T., G.F. Pinder and W.P. Saukin (1977). Modeling of leachate and soil interactions in an aquifer. Proceedings of the Third Annual Municipal Solid Waste Research Symposium on Management of Gas and Leachate in Landfills. EPA-600/9-77-026. pp. 95-103.

36. Appel, C.A.; Bredehoeft, J.D. (1978). Status of groundwater modeling in the U.S. Geological Survey. Washington, DC; U.S. Department of the Interior.

37. Conway, R.A., editor (1982). Environmental risk analysis for chemicals. Van Nostrand Reinhold Co., New York, NY.

RECEIVED April 15, 1983.

Modeling of Human Exposure to Airborne Toxic Materials

G. E. ANDERSON

Systems Applications, Inc., San Rafael, CA 94903

Under contract to the Systems and Strategy
Development Division of the OAQPS/EPA,
Systems Applications developed and applied
modeling methods for the estimation of
human exposure and dosage from airborne
materials. The model is intended for a
screening analysis of the impacts of
chemicals under EPA review as potentially
hazardous by the definitions of the NESHAPS
program.
   The analysis methods are national in
scope and address emissions from a wide
variety of industrial and community source
types. The materials reviewed are of
widely disparate natures. They include
metals, and bulk and trace hydrocarbons,
including chlorinated and oxide derivatives
of hydrocarbons. The analyses are intended
to be preliminary screening analyses for
use in scoping and prioritizing regulatory
attention to toxic exposures from the
chemicals studied.
   The modeling package, delivered to the
EPA, includes nationwide data bases for
emissions, dispersion meteorology, and
population patterns. These data are used
as input for a Gaussian plume model for
point sources and a box model for urban-
wide area sources. Prototype modeling is
used for point sources that are too
numerous to define individually. Building
wake effects and atmospheric chemical decay
are addressed.

0097–6156/83/0225–0067$06.00/0

The modeling approach has recently
been modified to a grid form so as to
address problems of exposure from multiple
sources and/or multiple chemicals
simultaneously.

Origins of Atmospheric Risk Analysis

Transfers of materials across tissue surfaces exposed to the
atmosphere are critical to life processes for humans, other
animals, and plants. Thus, living things are particularly
susceptible to harm by airborne irritants or toxins. The
risk of such harm has been a major motivation for the
development of techniques for the analysis of atmospheric
dispersion.
    Much of the initial development of Gaussian modeling
and definition of dispersion paramenters was done during and
after World War I in addressing the problem of poison gas
dispersal. These studies involved the definition of risk
factors, such as exposure and dose. The next intensive
development effort came during and after World War II with
the nuclear weapons program.
    Airborne poisons in the nuclear weapons progam were not
limited to radioactive materials released from weapons. The
weapons technology involved the use of many exotic
materials, some of which were toxic (e.g., beryllium).
Hazardous releases of these materials occurred in industrial
settings in urban areas and were studied by the Atomic
Energy Commission as occupational and public health
problems.
    Definitions and techniques of risk analysis for
atmospheric pollutants developed in these military and
derivative programs were described in depth in "Meteorology
and Atomic Energy" (1). At the time that the weapons-
related concerns were being codified, public concerns for
and governmental regulation of the nation's severely
deteriorated air quality was leading to the development of a
greatly expanded array of analysis techniques.

Elements of Atmospheric Risk Analysis

Although selection of the appropriate analysis techniques is
often very problem specific, the basic elements of human
health risk analysis are few, as presented in Figure 1. The
figure shows that the aggregate risk to human health from
exposure to an airborne pollutant results from two
factors: (1) the spread of the primary agent (and/or its
transformation products) from its source(s) to contact with
people, and (2) the characteristics of the agent's action on
the people who are exposed to it. The useful expression of

the risk depends on joint statistical considerations of
agent dispersion and characteristics of the human receptors.

Each of the main risk analysis elements consists of
three interactive studies. Exposure estimates result from
the integration of pollutant dispersion patterns and human
population patterns. The dispersion patterns, in turn,
result from the joint action of emissions and dispersion
processes.

The health effect side of the diagram shows that unit
risk estimates result from interactive analyses of health-
affecting processes in the human body and observed effects
in human populations (epidemiology). Health effects are
identified by integrating clinical studies on humans or
animals with studies of physical and chemical responses to
pollutant agents in the human body.

Weaknesses in any of the risk-structure "building
blocks" limit the credibility or usefulness of the risk
estimates. Conversely, each constituent analysis should,
most appropriately and efficiently, be of comparable rigor
and detail with regard to each other. Note, however, that
the results of each building block study are of value in
themselves.

This paper focuses on issues in the "dispersion" block
of Figure 1. These issues must be addressed, however, in
the context of a health effects problem. Some knowledge of
the health risk is necessary to properly scope the exposure
analysis.

A health risk for atmospheric pollutants is based on
the concept that adverse physiological changes may be
produced in human tissue that has contacted or absorbed some
airborne material. The change might depend--at least
statistically--on some characteristics of the individual
(e.g., age, sex, occupation, racial background), on the
complete time pattern of the pollutant received (amount of
dosage received over exposed time), and on any measure of
that pattern. (Exposure is the occurrence of contact
between human and pollutants. Dose is the total amount of
material received. In this paper the concentration to which
a person is exposed on an annual average basis is a measure
of the potential dose he may receive. Individual dose,
summed over all exposed persons, is referred to here as
dosage.) Pollutant patterns can be measured in several
ways: total dosage, dosage in a given time, exposure at or
above a given dose rate, and linear or nonlinear and
continuous or noncontinuous functions of any of the above
measures.

Appropriate methods of exposure analysis depend on the
form of the health effect function, which must be presumed
to depend on some function of the time history of
concentration to which a person is exposed. Even

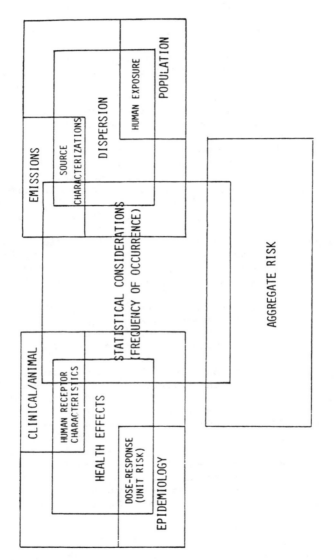

Figure 1. Relationship between the various dispersion studies, analysis product, and risk estimates.

statistical bases for identifying health effects functions
are usually weak; therefore, the health functions used in
practice typically are based on the simplest possible
measures of a concentration pattern. The two simplest
measures are exposure and dosage.

Exposure is generally related to a given concentration
level. This type of model is applicable to reversible
health effects. That is, below the "standard" concentration, the body can repair damage rapidly enough to
suppress symptoms. As the standard is exceeded, the human
body becomes less capable of repairing damage at a
satisfactory rate. Thus, damage symptoms appear in more and
more people. All of the so-called criteria pollutants
subject to National Ambient Air Quality Standards (NAAQS)
are presumed to produce reversible effects at NAAQS
levels. The NAAQS themselves are exposure types of
standards. Of course, sufficiently high exposure to such
pollutants can produce irreversible organ damage or death.

Some pollutants can produce irreversible cell or
genetic damage or irreversible cancerous consequences at
very low concentrations. Because of the irreversibility of
such effects, the total damage to the body can depend on the
accumulation of such events and thus on the pollutant dosage
that the body receives. If such damage is sustained at very
low levels, it may well be generally undetectable. Yet the
production of a cancer is a random event; it may occur on
the first "hit", but the probability of occurrence increases
with accumulated dose. Some bodily damage may be reversible
or inconsequential at low levels of exposure or dosage but
irreversible at higher levels. Such behavior is referred to
as "threshold" variation.

It is presumed that the effect of carcinogenic
materials is to produce critical cell damage. Thus,
carcinogenic health effects models generally are dose (i.e.,
integrated exposure) models, not exposure models. The lack
of firm statistical bases often leads to the adoption of
nonthreshold, linear models, even though thresholds and
nonlinear effects might be expected.

If linear (dose) models without thresholds are to be
used for carcinogen (or other) risk assessment, estimation
of exposure at specified levels becomes irrelevant to risk
assessment or, at least, its use is nonintuitive. For
example, a carcinogen risk analysis may be based on a
linear, nonthreshold health effects model. The total health
risk would thus be proportional to the long-term exposure
summed for all affected people for the identified period,
and exposure of many people at low concentrations would be
equivalent to exposure of a few to high concentrations. The
atmospheric dispersion that reduces concentrations would
also lead to exposure of more people; therefore, increments

to population risk would not necessarily diminish with
increasing dispersion time or distance. Limits to human
risk would exist only if the concentration or population
patterns were bounded, as for example, by either chemical
decay or scavenging by such phenomena as precipitation and
respiration.

Modeling Approaches in Use

The appropriate time and space scales are imposed by
estimated health effects functions, source and population
patterns, data quality and availability, and by the user's
information needs. These constraints have led to a wide
range of analytical approaches.

The simplest approach is to simply identify the
likelihood of contact between people and pollutant at
significant concentrations. This is often the extent of
"risk" analysis of preliminary, multi-media, problem-scoping
studies of hazardous or toxic materials (2). In the most
detailed approach, finely resolved spatial and temporal
patterns ("micro-environments") of concentration are
measured for each of many individuals representing finely
resolved population groups ("cohorts") characterized by
unique "activity patterns" (3, 4).

In many risk analyses standard dispersion models,
available from the EPA for regulatory compliance purposes,
are used to compute concentration patterns for prototypes of
a class of sources, and the patterns are convolved with
population patterns that are characteristic of the source
sites (5, 6). A similar level of analysis detail that
relies on measured pollutant (ozone) concentration in each
county of the Northeast Corridor rather than on modeled
concentrations was used by Johnson and Capel (7).

It is important in defining any analysis scheme that
the analysis elements be consistent in scope, scale, and
detail with each other and with the purposes of the
analysis. Thus details of cohort exposure in micro-
environments can provide valuable information on populations
at risk if, in fact, pollutant concentrations are functions
of micro-environments. It appears that micro-environments
are clearly important in carbon monoxide (CO) exposure
analysis because automobile generated CO concentrations are
highly correlated with automobile usage patterns. It is not
clear that ozone exposures are so correlated. Ozone
commonly exists in "clouds" that are large compared to any
one micro-environment, but drift over an area large compared
to their size in the course of their formation and decay.

Human Exposure Analysis at Systems Applications, Inc.

The Systems Applications Human Exposure and Dosage Model
(SHED) was developed under contract to the EPA Office of Air
Quality Planning and Standards. The work was done under the
OAQPS mandate to review chemicals in use for potential
regulation under the National Emissions Standards for
Hazardous Air Pollutants (NESHAPs). The large number of
chemicals identified as possibly carcinogenic left the
Pollutant Assessment Branch with the difficult task of
making "screening" estimates of the exposure/dosage of the
candidate pollutants so as to be able to order their
priorities for more detailed studies.

Thirty-five chemicals were specified for this initial
screening. The list of chemicals in Table I contains
materials of quite disparate character. Distinctive
characteristics include the following.

1. Phase--Solids, liquids, and gases (at ambient
    conditions) are represented.
2. Chemical Reactivity--Some are nonreactive; some decay
    by atmospheric chemical processes; and some are created
    by such processes.
3. Ubiquity-- Some are widely distributed; others are
    found in isolated locations, isolated times, or both.
4. Mode of emission--In general, when a pollutant is
    exposed to the atmosphere some fraction is lost to the
    atmosphere. Since each material is handled
    differently, it enters the air by a different mode.
    Some identified modes are: evaporation through a
    stack; emission through a vent (a vent is not designed
    to elevate the emitted material--a stack is); leaks in
    plumbing or storage containers; and wind-blown dust.
5. Emission rate--Rates range from minute to massive.
6. Proximity to people--Materials are emitted from sites
    of varying remoteness.

Because of the number of characteristics that must be
addressed, three different methods were used for estimating
concentration patterns, one method for each of three
categories of sources. The three source categories are as
follows.

Major, specific point sources. These consist of
individually identified sources, usually manufacturing
plants. Such sources have known locations and modes and
rates of emission. Each accounts for a significant fraction
of the national emissions of some species.

Other point sources. Sources that are too numerous,
small, or of uncertain location, and yet produce isolated
patterns of significant concentration, are not treated

Table I. List of Chemicals for Human Exposure/Dosage Estimation

No.	Chemical
1	Acetaldehyde
2	Acrolein
3	Allyl chloride
4	Benzyl chloride
5	Beryllium
6	Carbon tetrachloride
7	Chlorobenzene (mono)
8	Chloroform
9	Chloroprene
10	m-Cresol
11	o-Cresol
12	p-Cresol
13	o-Dichlorobenzene
14	p-Dichlorobenzene
15	Dimethylnitrosamine
16	2,3,7,8-TCDD (Dioxin)
17	Epichlorohydrin
18	Ethylene oxide
19	Formaldehyde
20	Hexachlorocyclopentadiene
21	Manganese
22	Methylene chloride
23	Nitrosomorpholine
24	Nickel
25	Nitrobenzene
26	PCBs
27	Phenol
28	Phosgene
29	Propylene oxide
30	Toluene
31	1,1,1-Trichloroethane
32	Trichloroethylene
33	m-Xylene
34	o-Xylene
35	p-Xylene

specifically. Rather, a prototype of such sources is
defined, and the results of prototype analysis are
multiplied by estimated numbers of sources that the
prototype represents. Degreasers are an example of sources
that were treated by prototype.

<u>Area Sources.</u> Sources that are so numerous and emit so
little that patterns of concentration are analyzed only en
masse. Such souces include both stationary (e.g., home
chimneys) and mobile (e.g., automobiles) types. Emission
rates per unit area are estimated; emission modes are not
addressed.

Emission rates, modes, locations, and times must be
described for each species studied. The emission work was
done by Hydroscience, Incorporated (HI), of Knoxville,
Tennessee. Emission characterization involved review of
trade literature, files of the various states, EPA reports
and data, and site visits and correspondence with staff of
specific sources.

The results of this program include the completion of
emissions summaries that identify source locations and
estimate the total nationwide emissions of the 35 chemicals.

Dispersion Modeling

The estimation of human exposure/dosage to atmospheric
concentrations of the studied chemicals involved three
computational tasks:

1. Estimation of annual average concentration patterns of
    each chemical in the region about each source.
2. Estimation of the population pattern over the area of
    each computed concentration pattern.
3. Computation of sums of products of the concentration
    and population patterns to provide exposure/dosage
    estimates.

Concentration Patterns

The large number of chemicals and sources that were modeled
in this program would consume large computer resources if
conventional modeling systems had been used. To keep
computer costs within reasonable bounds while ensuring that
the computing effort would meet program needs, we developed
a combined "reactive prototype" and "matrix" modeling
system.

The estimation of concentration patterns was done with
a different approach for each of the three source types
described above (specific point sources, prototype point
sources, and area sources). Each type of source requires a
different modeling approach. In addition, the concen-

trations of some of the selected chemicals depend on
reactions in large-scale plumes of photoreactive materials
from urban regions or industrial complexes.

Although Systems Applications has developed and used
many types of photochemical simulation models, application
of such models to the number and variety of sources studied
in the present program would require large labor and funding
resources; hence, these models were not recommended for this
program.

Major (Specific) Point Sources

Major sources of most of the selected chemicals are
specifically identified chemical manufacturing plants.
Concentration patterns due to unit emissions from such
sources depend most strongly on the following factors:

> Source elevation above terrain,
> Wind vectors (speed and direction), and
> Dispersive effects (intensity of atmospheric
> turbulence).

Long-term average concentrations depend on the time
histories of the meteorological parameters. A useful
simplification that greatly reduces computational
requirements is the computation of long-term average
concentrations by taking climatological weighted sums of
concentrations computed for a set of discrete states of the
atmosphere.

In the present study the computations were carried out
taking into account the following source-specific factors:

1. Climatological data from nearest or otherwise most
   appropriate recording station.
2. Individual treatment of releases from each identified
   process or vent within a plant.
3. Release height, speed, and buoyancy.
4. Effects of wakes from nearby structures.
5. Diurnal variations of emissions.
6. Seasonal variations of emissions.
7. Urban or rural character of area.
8. Atmospheric chemical reaction after release of
   emissions.

General Point Sources Represented by Prototype

Some point sources are not treated individually because of
their number and emissions strength; such sources are too
numerous, their emissions are too small to warrant
individual modeling, or both. Unlike area sources, these

sources are separated widely enough that their patterns of
pollution impact do not generally overlap. In lieu of the
individual modeling of each such source, a prototype source
is defined to represent each such source; dispersion and
exposure/dosage patterns are computed for the prototype; and
results are multiplied by the number of sources the
prototype represents.

Such sources were modeled using the matrix model for
unit emissions rates rather than emissions rates for actual,
specifically identified sources. When appropriate,
prototype sources were analyzed for each region of the
country using meteorological data representative of that
region. Nine geographic regions in the United States were
used (Figure 2); a model source was defined for each generic
source category in each analyzed region.

Area Sources

Area sources of either a selected chemical or a precursor
present a common problem for modeling. In particular, the
rich and complex patterns of hydrocarbon emissions from
general urban and industrial sources either include or might
produce through atmospheric photochemical reactions some of
the species on the analysis list. The treatment of such
species in photochemical airshed modeling is difficult (8,
9). The effort required for any one such exercise is
substantial, and the effort required for a comprehensive
analysis of all urban regions relevant to this program would
be prohibitive.

Reactive effects were treated through judicious scaling
of nonreactive results by factors developed by photochemical
"prototype" definitions. Nonreactive modeling of area
sources was carried out by use of a box model (10). This
type of model can be used to treat general, undifferentiated
source densities in an urban region. Box model results for
each wind speed and stability, weighted by climatological
probabilities, were used to compute long-term averages.

Basic box models cannot portray effects of nonuniform
source patterns. If, for particular chemical species or
particular source classes, the dependence of emissions on
population density or other identifiable parameters is
apparent and significant, we have used modifications to the
box modeling approach. As an example, it might be assumed
in modeling products of combustion of the lighter fuel oil
distillates that source distribution patterns are
proportional to population density patterns, because most of
such fuel is burned in residential furnaces in cold weather
cities.

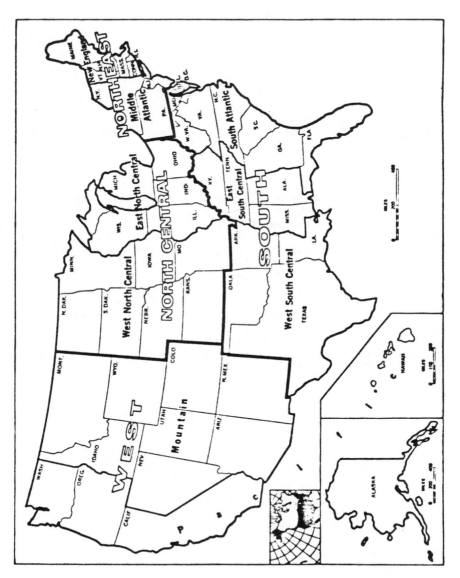

Figure 2. U.S. geographical regions.

Population Modeling

Population modeling was also done using different methods for each of the three types of sources: major, specific point sources; prototype point sources; and area sources.

For major point sources, site-specific population patterns were extracted from U.S. Census Bureau files using data at the Enumeration District/Block Group (ED/BG) level. These data provide the finest resolution of population patterns available. The data were scaled from 1970 to a base year of 1978 using county growth factors published by the Census Bureau. Interpolations of population and concentration patterns were used to develop patterns of exposure/dosage that were then summed to produce source-specific exposure/dosage totals.

The same dispersion procedures were used for modeling of other point sources; but since only prototype sources were addressed, population data were required only for prototypical conditions in each geographic region. Prototypical population was represented by the average population density in the urbanized areas of each region, since nearly all sources treated by prototype were located in urban areas.

For area sources, only city-average population density and area were used for each city so modeled.

Results

The emissions study identified and provided computations of the concentration, exposure, and dosage patterns for the following.

1. There were 311 major chemical manufacturing or consuming plants covered in this study. Because some major chemical plants were sources of more than one chemical, specific point source modeling was applied for 538 plants. Since there may be more than one source type in a plant, dispersion–dosage modeling was conducted for a total of 1819 individual point sources in this study.
2. There were 62 source categories involved in the prototype modeling, each modeled in nine regions. Hence, the prototype point source modeling was conducted for a total of 558 prototype sources.
3. Gaussian dispersion model computations were made for all "urbanized areas" (248) for each of the 77 area source categories, for a total of 19,096 runs.
4. Gaussian dispersion model computations were made for all other cities (243) with a population over 25,000 for each of the 77 area source categories, for a total of 18,711 runs.

5.   Box model computations were made for 150 cities with
     populations between 2,500 and 25,000 for each of the 77
     area source categories, for a total of 11,550 runs.

     In total, emission estimates and dispersion,
population, and exposure/dosage computations were made for
51,734 cases.

Regional Grid Model of Exposure/Dosage

Although the SHED model has proved valuable for screening
individual species being studied for carcinogenicity, the
EPA has an interest in evaluating the actual hazard to
carcinogens experienced by exposed populations. People are
actually exposed to many chemicals simultaneously and may be
exposed to concentrations that result from many sources
impacting the same area. To address these issues, SHED was
modified so that superimposed concentration patterns from
multiple sources could be computed and that provision be
made for weighting the computed concentrations for each
chemical so that the accumulated hazard (risk) from
multichemical doses could be computed.
     These changes were made by developing a grid model,
SHEAR. With a grid definition, relative locations of
sources could be preserved. The actual grid used was the
irregular set of ED/BG centroids. Although the centroid
grid points were located at irregular points, the use of
this grid preserved all of the spatial resolution inherent
in the population patterns, allowed preservation of SHED
interpolation algorithms, and produced the minimum loss of
information through excessive interpolations. The model
developed has the following features.

1.   The user-specified modeling region may exclude internal
     unpopulated areas (e.g., water areas).
2.   Stack height and plume rise are treated in source and
     meteorology-specific fashion.
3.   Prototype sources are assigned locations by randomized
     rules.
4.   Emissions of any source class can be specified to be a
     function of the meteorological condition, e.g., field
     spraying of insecticides occurs on low wind speed
     hours.
5.   ED/BG specific population is used.
6.   Computed parameters are: Concentration, exposure (to
     concentration); dosage; hazard (health risk to
     individual); exposure (to hazard); and risk (health
     risk to population).

7. Each computed parameter can be summed and tabulated for any subset of sources (single source, source type, chemical, all sources and chemicals).

The model was applied to a sample problem for the Beaumont, Texas/Lake Charles, Louisiana region. Computations were carried out for nine chemicals: beryllium, carbon tetrachloride, chloroform, dioxin, epichlorohydrin, ethylene oxide, formaldehyde, manganese, and trichloroethylene.

The application was a model test only and no attempt was made to acquire or validate site-specific source data. Specific and prototype point sources and area sources were included; most risk came from the nonspecific, hypothetical sources. Risk was computed using unvalidated, hypothetical unit risk factors. Sample graphical results are presented in Figures 3, 4, and 5.

In each figure, the regional UTM coordinates are given and excluded areas are shown. Heavy lines are county boundaries. The plotted sources are ED/BG centroids; their size is roughly proportional to the district's population. Coded isopleths of concentration or hazard are presented, and the code values are listed to the right.

The first trichloroethylene isopleths represent impact patterns from a single specific source (a chemical plant). Figure 4 shows trichloroethylene isopleths resulting from all sources, and Figure 5 presents isopleths of net hazard from all chemical sources in the region.

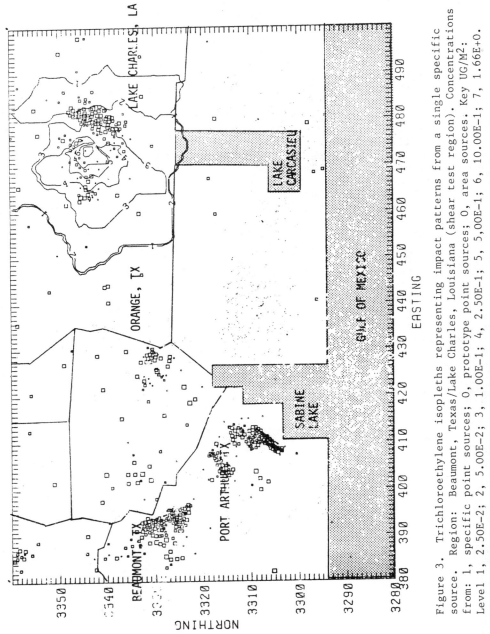

Figure 3. Trichloroethylene isopleths representing impact patterns from a single specific source. Region: Beaumont, Texas/Lake Charles, Louisiana (shear test region). Concentrations from: 1, specific point sources; O, prototype point sources; O, area sources. Key UG/M2: Level 1, 2.50E-2; 2, 5.00E-2; 3, 1.00E-1; 4, 2.50E-1; 5, 5.00E-1; 6, 10.00E-1; 7, 1.66E+0.

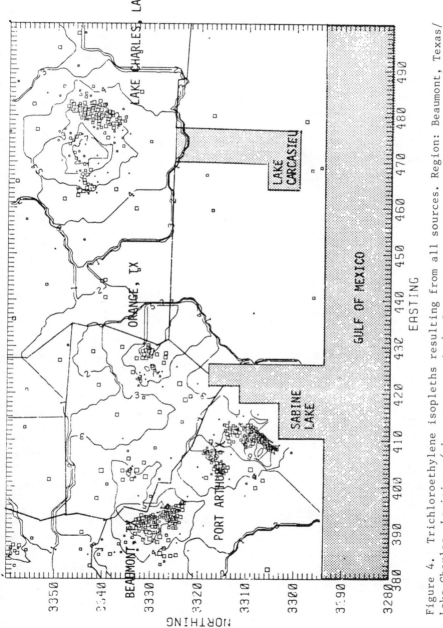

Figure 4. Trichloroethylene isopleths resulting from all sources. Region: Beaumont, Texas/
Lake Charles, Louisiana (shear test region). Concentrations from: 1, specific point sources;
3, prototype point sources; 1, area sources. Key UG/M², Level 1, 1.00E-2; 2, 2.50E-2;
3, 5.00E-2; 4, 10.00E-2; 5, 2.50E-1; 6, 5.00E-1; 7, 10.00E-1; 8, 2.50E+0; 9, 2.92E+0.

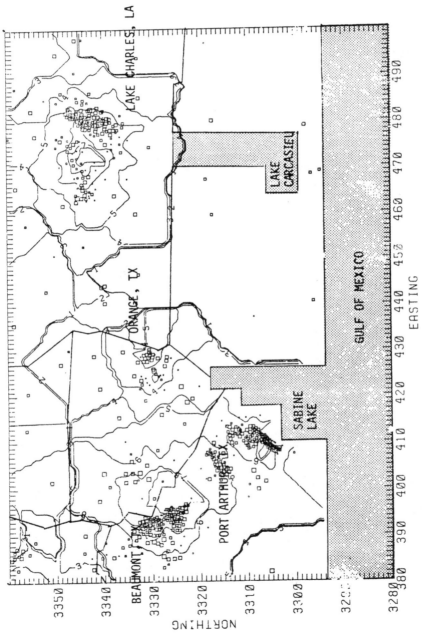

Figure 5. Isopleths of net hazard from all chemical sources. Region: Beaumont, Texas/ Lake Charles, Louisiana (shear test region). Key Hazard: Level 1, 5.00E-10; 2, 1.00E-9; 3, 2.50E-9; 4, 5.00E-9; 5, 1.00E-8; 6, 2.50E-8; 7, 5.00E-8; 8, 10.00E-8; 9, 1.75E-7.

Literature Cited

1. David H. Slade, "Meteorology and Atomic Energy 1968;" Air Resource Laboratories, Environmental Science Services Administration, U.S. Department of Commerce, 1968.
2. U.S. EPA, Discussion at EPA workshop on "Predictive Exposure Assessment," Washington, D.C. (unpublished), 6-7 April 1982.
3. W. R. Ott, "Models of Human Exposure to Air Pollution," Technical Report No. 32, Department of Statistics, Stanford University, Stanford, California, 1979.
4. T. Johnson and R. Paul, "The NAAQS Exposure Model (NEM) and Its Application to Nitrogen Dioxide," PEDCo Environmental, Inc., Durham, North Carolina, 1981.
5. K. R. Peterson and T. F. Harvey, "Meteorology and Demography Models for Risk Assessments of Accidental Atmospheric Releases of Nuclear Waste Phase 2 Methodology," Lawrence Livermore Laboratory, Berkeley, California, 1980.
6. B. E. Suta, "Assessment of Human Exposures to Atmospheric Benzene from Benzene Storage Tanks," Center for Resource and Environmental Systems Studies, Report No. 119, SRI International, Menlo Park, California, 1980.
7. T. Johnson and J. Capel, "Population Exposure to Ozone in the Northeast Corridor from May through October, 1979," PEDCo Environmental, Inc., Durham, North Carolina, 1980.
8. G. E. Anderson, S. R. Hayes, M. J. Hillyer, J. P. Killus, and P. V. Mundkur, "Air Quality in the Denver Metropolitan Region: 1974-2000," EPA-908/1-77-002, Systems Applications, Inc., San Rafael, California, 1977.
9. T. W. Tesche and C. S. Burton, "Simulated Impact of Alternative Emissions Control Strategies on Photochemical Oxidants in Los Angeles," RF78-22R, Systems Applications, Inc., San Rafael, California, 1978.
10. S. R. Hanna, "A Simple Dispersion Model for the Analysis of Chemically Reactive Pollutants," Atmos. Environ. 1973, 7, 803-817.

RECEIVED April 15, 1983.

MULTIMEDIA MODELS

The Role of Multimedia Fate Models in Chemical Risk Assessment

ALAN ESCHENROEDER

Arthur D. Little, Inc., Cambridge, MA 02140

This paper relates mathematical models for chemicals moving through air, water, soil and biota to methodologies for assessing health risks to individuals or ecosystems experiencing environmental exposures. The procedures for assessing risks are traced from sources to receptors, and the application of models to this process is described. The paper sets out to answer questions of how to select and link models in the context of risk assessment. The theory, structure, verification and application of the models themselves is left to other papers in this symposium. Acute risks are distinguished from chronic risks in the context of environmental regulatory requirements. A technique for selecting and assembling multimedia models based on release, environmental and receptor characteristics is described. The content of the paper is designed to unify other papers in the framework and organization of this symposium.

When chemicals are released in the environment, their hazard potential to human or ecological receptors depends upon the extent of contact between the receptors and the chemical. This exposure level is not only influenced by where, when and how much of the chemical is released, but also on its movement and changes in air, water, soil or biota relative to the locations of the receptors. Risk is defined as the probability of some adverse consequence in the health context, or as the probability times the extent of the consequence in the technology context. In this paper we shall examine and discuss how mathematical models are used to generate estimates of risk when more than one of the environmental media must be considered in tracing pathways connecting sources with receptors. The principal objective here is to place in perspective the

0097–6156/83/0225–0089$06.00/0

selection and application of fate models on the background of the needs perceived by government and industry for quantitative hazard or exposure analysis.

First, we investigate some of the regulatory motivations for chronic risk analysis. Next, it is necessary to point up the similarities and differences between acute and chronic risk and delineate the steps in estimating health risks posed by environmental chemicals. Following some illustrations of model structure, we conclude by discussing specific factors in fate analysis that suggest choices of model components.

Some Regulatory Background

Environmental control statutes and their administrative implementation through regulations have either implicitly or explicitly required chronic risk assessment. This has often been considered a yardstick in evaluating regulatory impact from a cost-effectiveness point of view. Indeed, in the closing weeks of the 97th Congress 2nd Session H.R. 6159 passed the House by voice vote and was pending in the Senate Commerce Committee at the time of this writing. The proposed legislation (Risk Analysis Research and Demonstration Act of 1982) establishes a program under the coordination of the Office of Science and Technology Policy (OSTP) for improving the use of risk analysis by those Federal agencies concerned with regulatory decisions related to the protection of human life, health and the environment. The bill would establish research, demonstration and coordination programs among these agencies. It further requires the Administrator of OSTP to present Congress with a plan for implementing risk analysis.

The Clean Air Act, the Federal Water Pollution Control Act, the Safe Drinking Water Act, the Federal Insecticide, Fungicide and Rodenticide Act, and the Toxic Substances Control Act implicitly require risk analysis in setting standards that the U.S. Environmental Protection Agency imposes through the state and local control agencies. (Indeed, the "zero discharge" goal in some statutes bypasses all needs for risk analysis.)

Highly toxic air pollutants fall under Section 112 of the Clean Air Act. Unlike criteria pollutants, these hazardous air pollutants must be controlled to protect the public health with an "ample margin of safety." Implied in this language is the belief in a discrete threshold of exposure below which no effects occur and from which a safety margin can be measured. Subsequent interpretations, however, indicated clearly that Congress did not equate safeguarding the public health with complete risk elimination.

The Federal Water Pollution Control Act (Clean Water Act) establishes nationally applicable effluent limitations using criteria based on different levels of control technology. For example, risk assessments were carried out in response to the settlement of litigation asserting a failure to set standards for 129 potentially toxic materials called "priority pollutants." The agreement stemming from this litigation was essentially ratified in the 1977 amendments to the Act and 1984 is the deadline year for the establishment of permissible effluent levels.

The risk methodology recommended for the water quality criteria involves either qualitative or quantitative estimates of concentrations of a pollutant in ambient waters which, when not exceeded, will "ensure a water quality sufficient to protect a specified water use." Criteria are intended for both the protection of human health and of ecosystems; however, they do not carry the authority of law.

The hazardous waste guidelines and regulations generated by the EPA in response to the Resource Conservation and Recovery Act of 1976 (PL 94-580) propose to cover methods of defining and identifying hazardous waste, standards for keeping records of containing and transporting these wastes, and standards for performance in the management of hazardous waste facilities, but do not explicitly require risk assessment.

Most of the provisions of the Toxic Substances Control Act (TSCA) of 1976 (PL 94-469) rely in some way on risk assessment of chemicals. Under the reporting requirements of the statute, any manufacturer, processor, or distributor of a chemical for commercial purposes must inform the EPA immediately after discovering any information which "reasonably supports the conclusion" that a chemical substance or mixture "presents a substantial risk of injury to health or to the environment" unless the EPA Administrator has been adequately informed already. EPA is mandated to establish regulations for testing new or existing substances when it is determined that there is not enough health or environmental information, that testing is necessary to develop such information and that the chemical or mixture "may present an unreasonable risk of injury to health or the environment."

Representations of adequate consideration of chronic risks are, therefore, necessary in the planning of many schemes for manufacturing, transporting, storing, use and disposal of potentially toxic waste materials. The combined effects of the statutes as described above have focused regulatory attention on the multimedia (air, water, soil and biota) aspects of such activities. It would appear as if the trend is toward acceptance of some risk rather than a guarantee (or hope) of complete safety of the public.

Chronic vs. Acute Risk Analysis

Environmental chemical releases due to human activities may be accidental (usually acute) or as an attendant consequence of some planned activity (usually chronic). Traditionally, spills have been separated from steady discharges because of statutory distinctions, but any integrated pollutant assessment must

consider both. Some materials are so hazardous that any routine
emissions are practically nil; therefore, inadvertent discharges
may dominate.

Space and time scales can be combined to draw the
distinctions between the risks due to these two types of
release. Acute risks are usually associated with immediate
effects of a release occurring within hours of the accident and
confined to within a few kilometers or less of its location.
Examples of this class of events are spills, fires, explosions
and their effects such as property damage, traumatic injury, or
sudden death.

Events that generate chronic risks may be the same as those
leading to acute effects or could be subtle releases distributed
over long periods of time. In either case, the term "chronic"
refers to longer term and potentially more widespread
consequences than those precipitated by acute risk events as
exemplified above. Whether caused by rapid or gradual releases,
chronic risks are occasioned by pollutant exposures of receptors
lasting days or even years. Some cases are difficult to
classify such as the short-term exposure that leads to an effect
which appears much later. Thus, the cause may be acute and the
effect, chronic. Their geographical ranges may extend over many
kilometers around an incineration site or along a transportation
corridor. The distributed use of potentially hazardous materials
such as pesticides generates chronic risk regardless of
geographical range.

Any analysis of risk should recognize these distinctions in
all of their essential features. A typical approach to acute
risk separates the stochastic nature of discrete causal events
from the deterministic consequences which are treated using
engineering methods such as mathematical models. Another tool
if risk analysis is a risk profile that graphs the probability
of occurrence versus the severity of the consequences (e.g.,
probability, of a fish dying or probability of a person
contracting liver cancer; either as a result of exposure to a
specified environmental contaminant). In a way, this profile
shows the functional relationship between the probabilistic and
the deterministic parts of the problem by showing probability
versus consequences.

Let us now turn our attention to the main steps of any
procedure constructed to anticipate or respond to the risk
analysis requirements set forth by the statutes reviewed above
or voluntarily established as product standards by industries.
It is important to note that this type of procedure is a
technical means to arrive at a quantitative estimate. The
decisions regarding the acceptability of the result is
sociopolitical and is, therefore, beyond the scope of this
discussion.

Components of a Chronic Risk Assessment

Materials Balance Analysis. The first step in our methodology is the establishment of flows of hazardous pollutants and their distributions among the environmental compartments. The time phasing of releases must be considered; for example, some releases may be instantaneous at a frequency of once a month while others may be continuous with seasonal or diurnal variations superimposed. Furthermore, the ultimate chronic risk will also depend upon the spatial disposition of releases; for example, a moving elevated point source will give ambient concentration patterns different from those from a stationary surface-based area source. Chemical speciation also must enter our materials balance description in some cases. A case in point is hexavalent chromium which has a higher order of carcinogenicity than trivalent chromium, which is also found in nature. Finally, partitioning among the media is an essential ingredient in the characterization of emissions or discharges - i.e., how much of a release enters the air, water, soil or biota? Or, put another way, into what compartment is an environmental release deposited initially? Answering this question sometimes involves skipping ahead to a short-term chemical fate analysis such as for a sudden spill, depending on material properties, fractions of the spilled material may be found in any or all of the four compartments (air, water, soil, biota) or at their interfaces. This is an example of how the output of an acute risk analysis can provide input to the chronic risk analysis by providing the instantaneous distribution of the released substance among the compartments.

Whether the releases to the environment are sudden or gradual, it is necessary to devise a systematic method to account for each component. One approach to this problem is based on a matrix having rows consisting of activities or sources (e.g., extraction, processing, manufacturing, storage, transportation, use, disposal and reclamation) and columns representing the media, air, earth, water and biota. Each non-zero element of this matrix array is filled out with the spatial, temporal and chemical detail called for above. The materials balance thus derived provides points of entry into the pathways of exposure that ultimately form the basis of the chronic risk assessments. Brown and Bomberger have discussed the methodology for this step extensively in another paper in this symposium (1).

Environmental Fate. Having characterized the entry of materials into the environment, we move into the second step of our procedure. The goal at this stage of analysis is to define ambient concentration of the material or its products in areas of concern for receptor (e.g., people, materials or ecosystem components) exposure. A family of computer simulation models has been developed for calculating the ambient levels of a material subjected to the simultaneous influences of transport, diffusion and transformation in a multimedia setting. This has been implemented by linking single medium models at the interfacial boundaries (such as the linking of an air model to a soil model by deposition and volatilization processes). These capabilities have grown intensively over the past five years largely due to the sponsorship of government and industry.

Examples of the need for multimedia models are found in contemporary problem areas. Polynuclear aromatic hydrocarbons and metals are emitted into the atmosphere as trace impurities with the products of coal combustion. The organics have low vapor pressure and partially condense on emitted particulates in a stack plume. The particulates are transferred to the soil by dry deposition, rainout or washout. The metals manifest themselves in relatively refractory oxides formed selectively among the fine size ranges of flyash particles. Both particle bound pollutants must be treated in chronic inhalation risk studies by a sequence of air dispersion and surface deposition processes, whereas the vapor fraction of organics remains in the air. Thus, the gas phase, aerosol and soil components are treated simultaneously in multimedia model studies.

Another case of multimedia fate modeling may be exemplified by human inhalation exposure estimates for PCB spills. The spill size is estimated considering both spread and soil infiltration. Volatilization calculations were carried out to get transfer rates into the air compartment. Finally, plume calculations using local meteorological statistics produced ambient concentration patterns which can be subsequently folded together with population distributions to obtain exposures.

Numerous examples of fate models are reviewed in other papers in this symposium. For example, single media models are covered for air by Anderson (2), for water by Burns (3), and for soil and groundwater by Bonazountas (4).

Receptor Exposure. Exposure modeling should produce a statistically representative profile of pollutant intake by a set of receptors. This is done by combining the space/time distribution of pollutant concentrations with that of receptor populations (whether they be people, fish, ducks or property made of some material that is vulnerable to pollutant damage). The accuracy and resolution of the exposure estimates are chosen to be consistent with the main purposes of decision making. These purposes include the following:

o   Screening of pollutants or sources to set priorities;
o   Evaluation of legislation or rulemaking;
o   Comparison of alternate ambient standards;
o   Planning of facilities at specific sites;
o   Support of field research programs; and
o   Design of real-time episode control systems.

It is clear that these goals place widely differing requirements on both the resolution and the accuracy of exposure estimates; thus, the approach selected should be optimized to fit the requirements.

The exposure models are designed to be compatible with fate analysis outputs whether they be for air, water or soil. If there are significantly different exposure or dose/response characteristics among various subpopulations, we form cohort groups differentiated by age, sex, occupation, level of activity or geographical habits. Limitations of available data derived from clinical, epidemiological and toxicological studies usually preclude distinction of dose/response curves among the cohorts; however, there often are sufficient data on indoor vs. outdoor levels, geographical variation, and occupational surroundings to allow some distinctions to be drawn among cohorts. Thus, peculiarity of microenvironments lead to differing exposure levels.

Because the significance of exposure has only been considered over the past few years, there is not as wide a selection of exposure models available as that for fate models. The latter have been applied for several decades to the calculation of ambient exposure levels compared with some standard values. Papers illustrative of human exposure assessments in this symposium include one on airborne pollutant exposure assessments by Anderson (2), a generic approach to estimating exposure in risk studies by Fiksel (5), and a derivation of pollutant limit values in soil or water based on acceptable doses to humans by Rosenblatt, Small and Kainz (6).

Risk Estimation. As mentioned above, chronic risk is expressed as a probability of occurrence per year or per lifetime of some adverse consequence caused by exposure to the pollutant. Statutory mandates have focused on human health effects as the primary expression of chronic risks. The basis of the risk calculation is the dose/response curve that relates the adverse effect to the amount or rate of a chemical taken in to the subject. Because of regulatory emphasis of cancer, most of the work devoted to the deviation of dose/response curves has been concerned with the probability of appearance of a tumor as the adverse effect.

The risk estimation procedure may be thought of as performing three distinct functions:

1. Conversion of experimental dose/response data into a form suitable for extrapolation of human risk using least squares or, more usually, maximum likelihood curve fits.
2. Generation of alternative dose/response models for risk estimation to emphasize the range of results generated by widely differing assumptions.
3. Display of risk levels for various subpopulations under various applications of technological or regulatory control of releases into the environment in order to relate social costs to risk reduction.

The dose/response models are intended to extrapolate both
from test animals to humans and from high doses to low. The
user should be concerned with assumptions underlying these
models. A range of assumptions for risk evaluation might be
obtained by making three choices of extrapolation formulas: the
one-hit linear model, the multistage model and the log probit
model. The one-hit formulation is based on a postulate that the
invasion of a cell by a single pollutant molecule can initiate a
tumor. This gives a straight-line relationship between dose and
response. The multistage models depend on a mechanism involving
multiple processes at various stages of cell division to cause a
tumor. Going from high doses to low doses, the multistage risk
drops off more sharply than the one-hit risk as dose is
decreased. At intermediate to low doses, however, multistage
asymptotically approaches linearity at some constant factor
placing it somewhat below one-hit. Log probit is a model based
empirically on a sigmoid-shape assumption for all dose/response
curves. This shape approximates the notion of a threshold;
i.e., a dose below which defense mechanisms, metabolism or
elimination processes intervene to prevent tumor formation.

All of the extrapolation models are predicated on the
supposition that there are no interspecies differences. None
assumes any synergism or antagonism with other pollutants, and
all of them scale effects by surface area in order to consider
the size of the receptor organism. No distribution is made
among the various entry routes into the body since the
pharmacokinetics, which describe the chemical's fate in the
organism, are not differentiated. Despite these limitations,
regulatory agencies use dose/response extrapolation for decision
making; therefore, the analyst must be mindful of the wide range
of values yielded by the various models at low dose and be aware
of the uncertainty of the risk results. Because of these
difficulties, it is often useful to stop at the exposure
calculations and compare exposure statistics with ranges of
values accepted and experienced in everyday life.

In this symposium a comprehensive overview of the risk
estimation step and its relationship to the output of multimedia
fate models is given in the paper by Fiksel (5). Examples of
the application of and linkage among the various techniques are
also presented in that paper.

Multimedia Model Characteristics

Model Types. If it is determined that exposure pathways of
interest intersect more than one of the media, the analyst is
faced with the need to link together single media models (or to
apply existing multimedia models). Despite claims to the
contrary, there is probably no single model that is appropriate
to all problems. Thus, a hybrid combination of boundary

conditions, algorithms and output displays is assembled to respond to specific problem needs. The requirements are expressed in terms of factors such as the following:

o   accuracy as evidenced by validation tests
o   time/space resolution
o   overall interval or spatial scale
o   resource availability (e.g., data processing system capability, user personnel level, budget and turnaround time)

These requirements are driven by the application whether it be product design, regulatory mandates, regional planning, standard setting, legislative drafting or control strategy design. The techniques available for multimedia modeling up to around 1978 were reviewed in a previous paper (7); this symposium is intended to provide the fundamentals and applications reflective of development efforts as well as the current state-of-the-art. Much in the way of useful background material is summarized in the proceedings of a workshop convened by the U.S. Environmental Protection Agency (8).

Beyond the stage of development described in these documents, multimedia model designs can be roughly categorized as either well-mixed compartment types or transport types. Either type may or may not handle chemical transformations. The principal distinguishing characteristic of a multimedia model is its capability to calculate flows across media boundaries. The content of the well-mixed compartment model is mostly concerned with boundary processes since spatial uniformity is assumed in each medium or phase. The parallel developments of Mackay's approach (9) and that reported by Neely and Blau (10) are examples of well-mixed compartment models. The most rudimentary form of Mackay's approach uses the thermodynamic equilibrium scenario by defining a set of fugacities whose evaluation determines the partitioning of a chemical among the media. Higher levels allow for steady flow and unsteady flow behavior in the compartments, but the key element in applying any of the well-mixed compartment models is estimation of compartment volume. This step inherently presumes some estimate of transport. The approach of Neely is laboratory-based and involves the use of a decision tree to select calculation algorithms. The two methods were comparatively analyzed by Lyman (11) who concluded that for single component organic chemicals both models are easy to use with a minimum of data and can be executed on a hand-held calculator. Considering the severe limitations on these models, they are useful for screening approximations.

The transport type of model becomes necessary where site-specific predictive cability is needed. Mathematically this type is distinguished from the well-mixed compartment by its dependence upon partial differential equations generated by

the space/time equations governing chemical composition.* In
general, these equations have variable coefficients and can be
nonlinear. This almost certainly requires the use of at least a
minicomputer if not a main frame system. The UTM model family
(8) exemplifies the transport type approach by linking a series
of single media simulations.
     One weakness of some multimedia models that must be
considered by the user is inconsistency of time scales.   For
example, if we employ monthly averaged air concentrations to get
rainout values on fifteen-minute interval inputs to a watershed
model, large errors can obviously occur. The air-land-water-
simulation (ALWAS) developed by Tucker and co-workers (12)
overcomes this limitation by allowing for sequential air quality
outputs to provide deposition data to drive a soil model. This
in turn is coupled to a surface water model.

     Current Examples. In this symposium the characteristics,
types and applications of multimedia models are exemplified.
The compartment type is reviewed in a paper by Mackay and
Paterson (13).   The fugacity approach is discussed and
applications are described for polychlorinated biphenyls in the
Great Lakes region. Applications of compartment modeling to
organic chemicals are covered by McCall, Swann and Laskowski
(14). The implementation of this type of approach using
laboratory data based properties estimates is illustrated.   The
key role of interfacial transport in compartment or transport
models is the focus of a paper by Bomberger and co-workers (15).
The combined influences of chemistry, phase change and
biotransformation are processes modeled at the
terrestrial-atmospheric interface.
     Another compartmental partitioning issue of major
consequence for pesticides is the dissolved versus adsorbed
fraction in an aqueous environment. Carter and Suffet (16)
present measurements of binding of pesticides to dissolved
fulvic acids that will provide inputs to compartment models.
Data from laboratory measurements used in compartment models can
often bypass costly field experiments in the screening stage.
Thomas, Spillner and Takahashi (17) have related the soil
mobility of alachlor, butylate and metachlor to physicochemical
properties of these compounds.
     In the area of transport-type models, soil/water systems
have been a primary area of development. The Hydrologic
Simulation Program (18) described in the paper by Johanson
simulates chemical movement and transformation in runoff,
groundwater and surface water in contact with soil or sediments.

---

*Strictly speaking, finite difference or finite element
 solutions to differential equations are simply multiplying the
 number of comparments many times, but the mathematical rules
 for linking cells in difference calculations are rigorously set
 by the form of the equations.

An application of transport and compartment-type models to hazard analysis is described in the paper by Honeycutt and Ballantine (19). The compound CGA-72662 running off from agricultural areas into surface waters was modeled in order to set safe application procedures consistent with the protection of aquatic environments. Patterson, et al (20) have adapted the UTM model to a software package that is generally applicable to fate assessments of toxic substances in air, water, soil and biota. Their work, now in working draft form, is being used by Dr. William Wood and Dr. Joan Lefler in the Office of Toxic Substances of the U.S. Environmental Protection Agency.

Despite the intensive efforts devoted to making new multimedia models, it seems as if relatively little attention is given to their verification through field or laboratory measurements. (One notable exception is US EPA's pesticide evaluation project sponsored by its Athens, GA laboratory.) The philosophy of model validation and the conclusions from testing programs are reviewed by Donigian (21) in his paper. Although the work described is mainly concerned with aquatic simulations, the need for carefully designed evaluation studies will continue to grow for multimedia models proposed for use in risk assessments.

Selection and Application of Model Components

Influence of Entry Modes of Pollutants into the Environment. In selecting an appropriate multimedia model, the user must begin by identifying several features that characterize the emission, discharge or release of the pollutant of interest into the environment. Following this identification, quantitative estimates of rates and distributions must be developed because the ultimate use of a set of fate models is to calculate ambient levels in terms of release rates, pollutant properties and environmental scenarios. Since the designation of pathways is the primary step in establishing the factors determining model selection, the process begins with a set of initial points for the candidate pathways. Choice of model structure depends on release points and on the main aspects of fate processes influencing the movement along each pathway. A formal description of this approach is in preparation by Bonazountas and Fiksel (22). Their handbook/catalogue will provide users with a direct and simple way to select appropriate models. It will supply background of the physical, chemical and biological issues that must be considered in fitting model characteristics to problem needs.

Releases into the environment may be natural or anthropogenic. Bacterial or mineral action may constitute worldwide generation sources that function independently of any human activity. If pollutant impacts are to be evaluated, both these sources and natural sinks, such as the oceans, soil

surface or atmospheric photolysis must be considered along with
the anthropogenic materials balance.

Owing to their relatively good level of predictability,
chronic emissions or discharges can be classified and
quantitatively characterized in a unified manner using a matrix
or tabular form as described previously. For a geographic
study, spatial differentiation can be introduced, for example,
by subdividing water into particular stream reaches, ponds,
aquifers, etc. A broad classification of activity categories
should transcend the usual inventory of stacks or discharge
pipes; it should include extraction, processing, manufacturing,
storage, transportation, use, disposal and reclamation. Again,
the categories can be refined by storage at manufacturing site,
storage at forwarding terminal, storage at distribution depot
and storage by users, to cite one example.

For acute releases, the fault tree analysis is a convenient
tool for organizing the quantitative data needed for model
selection and implementation. The fault tree represents a
heirarchy of events that precede the release of concern. This
heirarchy grows like the branches of a tree as we track back
through one cause built upon another (hence the name, "fault
tree"). Each level of the tree identifies each antecedent
event, and the branches are characterized by probabilities
attached to each causal link in the sequence. The model
applications are needed to describe the environmental
consequences of each type of impulsive release of pollutants.
Thus, combining the probability of each event with its
quantitative consequences supplied by the model, one is led to
the expected value of ambient concentrations in the environment.
This distribution, in turn, can be used to generate a profile of
exposure and risk.

If required by the model(s) to be used, back-up data for
each entry in the matrix or table may be supplied to resolve the
total mass flow into spatial cells (UTM coordinates, depth or
height), temporal cells (hourly frequency distributions, diurnal
cycles, seasonal subdivisions or secular trends on annual
intervals) or speciation cells (by valency state of anions or by
hydrocarbon structure, for example). The level of difficulty
encountered by the user in supplying these data may influence
the choice of model(s).

Dynamics of Chemicals in the Environment. In identifying
pathways and, hence, models, the user must also consider what
becomes of the pollutant as it enters the environment. The
dominance of various factors over others will determine both
pathway selection and model selection in an integrated pollutant
assessment.

Within any medium of the environment, three types of
process (defined here as intramedia processes) govern the
pollutant concentration at each point at each time:

o   Advection – mass movement of the medium carrying the material along.

o   Diffusion – movement or spread of the pollutant as relative to the mass of the medium as driven by molecular or turbulence scale dynamics.

o   Transformation – production or consumption of the pollutant usually driven by chemical reactions.

Superimposed on these mechanisms of change operating in the bulk volume of each medium are processes that transfer the pollutant from one medium to another. Some conceptual model frameworks lump intermedia transfers together with embedded transformation processes* causing unnecessary mathematical confusion of boundary value specifications with source term formalisms. Examples of intermedia pollutant transfers are as follows:

o   Surface deposition (rainout, washout, fallout and dry)

o   Evaporation (codistillation or volatilization)

o   Adsorption - desorption

In choosing a model, the user can optimize fate assessment efforts by delineating first, the source release patterns and second, the dominant dynamical processes. Taking the intramedia processes first, one can address model criteria by considering the ratio of characteristic times. The advection time is the principal length scale of the domain L divided by the average flow speed u; i.e.

$$\tau_d \sim L/u$$

Typically, L may be stream reach distance and u, flow velocity. The diffusion time is approximated by the random walk hypothesis and is approximated by:

$$\tau_d \sim \Delta^2/2D$$

where Δ is the characteristic transverse direction (e.g. stream depth) and D, the transverse diffusivity, be it turbulent or molecular. Finally, the transformation time is approximated by

$$\tau_t \sim C/\overset{\circ}{C}_t$$

---

*Source terms, for example, are sometimes written in the equation separately for chemical production sources and for emission sources.

where $\overset{\circ}{C}_t$ is the average rate of concentration change due only to transformation (typically a chemical reaction rate) and C, the average concentration in the domain.

Let us examine three examples of how these times are used in model selection. If $\tau_t \ll \tau_a$ and $\tau_t \ll \tau_d$, there is rapid chemical change before any movement occurs. If $\tau_t \gg \tau_a$ and $\tau_d \ll \tau_a$, there is little chemical change and diffusion spreads the pollutant rapidly so that the mixture is homogeneous. If $\tau_t \sim \tau_d \sim \tau_a$, all processes act simultaneously. Taking these cases in order, we see that the first case is trivial requiring no model (except possibly a reacting plume in the near field). The second case is approximated by a nonreactive box model and the third, by a full reactive diffusion model.

Source geometry, interphase transfer and time dependencies must be superimposed on the above features to aid the user in choosing a set of models. For example, the advection distances would be different for point and area sources. Also, the significance of source location must be considered in light of interphase transfer efficiency; e.g., water discharge of a high volatility, low solubility material transfers the problem immediately from one of water modeling to air modeling. Implicit in these environmental dynamic considerations are three principles that may help guide the catalogue user:

- o   Intramedia processes are largely assessed on the basis of environmental scenarios
- o   Intermedia transfers are largely determined by the pollutant's fate properties
- o   Chemical transformations can figure in both of the above

Clearly, we can find exceptions to these rules: molecular diffusivity is a pollutant fate property, but may control an intramedia process; rainfall history is an environmental scenario characteristic, but may control an intermedia transfer.

In summary then, one should analyze the problem at systems level prior to model selection based on entry characteristics and environmental dynamics of the pollutant. Experience suggests that it is better to rely on intuition and a few calculations than to construct a formal logical decision tree for guiding this process. Often, the compartment screening models are helpful at this stage. Characterization of the sources, the environment and the fate properties is an essential prerequisite to any procedure.

Concluding Remarks

We note that multimedia fate modeling constitutes a central link in the chain of calculations forming a risk analysis. Although regulatory mandates, as described above, have constituted the primary motivation for fate modeling in hazard assessment, sound risk management practices will provide further impetus in the future as more anticipatory activity becomes a part of corporate planning. Multimedia models are being developed for various specific applications, and components are becoming available for the assembly of custom designed hybrid models. A decision procedure is evolving for selecting appropriate components.

The variety and depth of the papers presented at this symposium are ample evidence of the high level of interest now focused on this field. Taken as a whole, these papers not only provide a record of where we stand, but also provide a textbook for potential model users.

Literature Cited

1. Brown, S.L. and Bomberger, D.C., Chapter 1 in this book.
2. Anderson, G.E., Chapter 4 in this book.
3. Burns, L., Chapter 2 in this book.
4. Bonazountas, M., Chapter 3 in this book.
5. Fiksel, J.R., Chapter 15 in this book.
6. Rosenblatt, D.H., Small, M.J. and Kainz, R.J., Chapter 14 in this book.
7. Eschenroeder, A., Toxic Substances Journal 4, 38 (1982).
8. Haque, R. (ed.), Proc. of the Workshop on Transport and Fate of Toxic Chemicals in the Environment, 1981.
9. Mackay, D. Environmental Sci. and Tech. 13, 1218 (1979).
10. Neely, W.B. and Blau, G.E., "Pesticides in the Aquatic Environment," Plenum 1977.
11. Lyman, W., Prediction of Chemical Partitioning in the Environment, U.S. Environmental Protection Agency, Final Draft Report, May 1981.
12. Tucker, W.A., Eschenroeder, A.Q., and Magil, G.C., "Air Land Water Analysis System (ALWAS): A Multimedia Model for Assessing the Effect of Airborne Toxic Substances on Surface Water Quality," Arthur D. Little, Inc. report to EPA, Contract G8-03-2898, 1981 (in preparation; soon to be available from NTIS or from W.A. Tucker in abbreviated form as "The Air, Land, Water Analysis System (ALWAS): Multimedia Pathways of Organic Pollutants in Freshwater Systems," American Society for Civil Engineers National Convention, Las Vegas, Nevada, April 26-30, 1982).
13. Mackay, D. and Paterson, S., Chapter 9 in this book.
14. McCall, P.J., Swann, R.L. and Laskowski, D.A., Chapter 6 in this book.
15. Bomberger, D. C., Gwinn, J. L., Mabey, W. R., Tuse, D. and Chou, T-W., Chapter 10 in this book.
16. Carter, C.W. and Suffet, I.H., Chapter 11 in this book. Thomas, V.M., Spillner, C.J., and Takahashi, D.G.Scher,
17. H.B., Chapter 12 in this book.
18. Johanson, R.C., Chapter 7 in this book.

19. Honeycutt, R.C. and Ballantine, L.G., Chapter 13 in this book.
20. Patterson, M.R., Sworski, T.J., Sjoreen, A.L., Browman, M.G., Contant, C.C. Hetrick, D.M., Murphy, B.D. and
21. Raridon, R.J., A User's Manual for UTM-TOX, a Unified Transport Model, Oak Ridge National Laboratory Report 1AG No. AD-89-F-1-399-0 (in review).
    Donigian, A.S., Jr., Chapter 8 in this book.
    Bonazountas, M. and Fiksel, J.R., Enviro/HC: Environmental
22. Mathematical Handbook/Catalogue, Arthur D. Little, Inc., Report to the Environmental Protection Agency (in preparation).

RECEIVED April 15, 1983.

Partition Models for Equilibrium Distribution of Chemicals in Environmental Compartments

P. J. McCALL, R. L. SWANN, and D. A. LASKOWSKI

Dow Chemical, Midland, MI 48640

Distribution of organic chemicals among environmental compartments can be defined in terms of simple equilibrium expressions. Partition coefficients between water and air, water and soil, and water and biota can be combined to construct model environments which can provide a framework for preliminary evaluation of expected environmental behavior. This approach is particularly useful when little data is available since partition coefficients can be estimated with reasonable accuracy from correlations between properties. In addition to identifying those environmental compartments in which a chemical is likely to reside, which can aid in directing future research, these types of models can provide a base for more elaborate kinetic models.

Increased production and use of chemicals have created a need to better understand the fate and effects of chemicals in the environment. Recognition of environmental concerns by regulatory agencies has led to new legislation aimed at finding answers to important questions regarding the distribution and behavior of chemicals in the environment. Historically laboratory tests have investigated individual process associated with movement (soil leaching, volatility, adsorption, etc.) and transformation (soil degradation, hydrolysis, photolysis, etc.) of chemicals. The current thrust in environmental chemistry is to integrate environmentally meaningful laboratory data into a realistic description of the "real world" behavior of a chemical. The underlying goal of this research is to reach in the most efficient manner and in the shortest time possible, a reliable assessment of fate. Several approaches have evolved which, in general, can be described as models of the environment or parts of the environment. Models generally fall into two categories; physical or mathematical. Physical models often termed microcosms or model ecosystems attempt to isolate a representative segment of the

0097–6156/83/0225–0105$06.00/0

environment within which the fate of a chemical is observed in
order to describe its fate. Mathematical models attempt to define
all the important processes which act on a chemical in the environment and incorporate them into a description of chemical behavior as a function of the variables acting on the system. The type
of model approach taken depends on the degree of accuracy and precision expected and the questions asked of the model.

Two classes of mathematical models have been developed: those
which are specific and attempt to describe the transport and degradation of a chemical in a particular situation; and those which
are general or "evaluative" and attempt to generally classify the
behavior of chemicals in a hypothetical environment. The type of
modeling discussed here, equilibrium partitioning models, fall
into the latter category. Such models attempt, with a minimum of
information, to predict expected environmental distribution patterns of a compound and thereby identify which environmental compartments will be of primary concern.

Partitioning Models

In its simplest form a partitioning model evaluates the distribution of a chemical between environmental compartments based
on the thermodynamics of the system. The chemical will interact
with its environment and tend to reach an equilibrium state among
compartments. Hamaker(1) first used such an approach in attempting to calculate the percent of a chemical in the soil air in an
air, water, solids soil system. The relationships between compartments were chemical equilibrium constants between the water
and soil (soil partition coefficient) and between the water and
air (Henry's Law constant). This model, as is true with all
models of this type, assumes that all compartments are well mixed,
at equilibrium, and are homogeneous. At this level the rates of
movement between compartments and degradation rates within compartments are not considered.

Mackay(2,3) building upon the earlier work of Baughman and
Lassiter(4) advanced the development of partitioning modeling
using the concept of fugacity. Here chemical equilibrium or partition coefficients between two phases are expressed as an "escaping tendency" the chemical exerts from any given phase. Thus,
when a system is at equilibrium the fugacity in each compartment
matches that in any other compartment. Fugacity has the units of
pressure and partition coefficients between two phases are the
ratio of the fugacity capacities in each phase.

In Mackay's development of an equilibrium model a slice of
the earth is selected as a unit world or model ecosystem. Fugacities are calculated for each compartment of the ecosystem and the
overall distribution patterns of a given chemical are predicted.
In a similar approach McCall et al.(5) have defined a model ecosystem which represents a unit world, however, this development
incorporates standard chemical equilibrium expressions into a

series of simultaneous equations to predict distribution. Mathematically both approaches are essentially the same with the exception that different units are used. The end result is the same. A model is obtained which predicts distribution patterns of chemicals in a simulated environment representative of a segment of the world. The goal is not to predict actual expected environmental concentrations, but to predict expected behavior. To which phase is the substance likely to migrate; will a pesticide applied to soil leach or be volatile; will a chemical accumulate in the biotic compartment; and so on.

The method of using fugacity calculations will be discussed later in this symposium, therefore a detailed description will not be given in this paper. The description of equilibrium models using chemical equilibrium expressions will be discussed with the recognition that the two approaches are very much the same.

Environmental Partition Coefficients

Soil Sorption Constant – Soil/Water (K_{OC}). The distribution of a chemical between soil and water can be described with an equilibrium expression that relates the amount of chemical sorbed to soil or sediment to the amount in the water at equilibrium.

$$Kd = \frac{\mu g \ chemical/g \ soil}{\mu g \ chemical/g \ water} \qquad (1)$$

where

Kd = sorption coefficient
μg chemical/g soil = concentration of adsorbed chemical
μg chemical/g water = concentration of chemical in solution

The primary active surface that interacts with the chemical in the sorption process has been shown to be the organic fraction of the soil(6-10). Therefore, the sorption characteristics of a chemical can be normalized to obtain sorption constant based on organic carbon (K_{oc}) which is essentially independent of any soil.

$$K_{oc} = \frac{\mu g \ chemical/g \ organic \ carbon}{\mu g \ chemical/g \ water} \qquad (2)$$

This value, like other partition coefficients is a measure of the hydrophobicity of a chemical. The more highly sorbed, the more hydrophobic a substance is.

Henry's Law Constant – Water/Air (K_w). The distribution of a chemical between water and air is an expression of Henry's Law which can be written as follows(11).

$$K_w = \frac{C_{water}}{C_{air}} = \frac{T(WS)}{16.04 \ PM} = \frac{1}{H} \qquad (3)$$

where

K_w = recipricol of Henry's Law Constant (H)

C_{air} = concentration of the chemical in air (mg/liter)

C_{water} = concentration of the chemical in water (mg/liter)

P   = vapor pressure of pure chemical (torr)
M   = molecular weight
T   = temperature (°K)
WS   = water solubility (mg/l)

Such a relationship describes how a chemical will partition between water and the atmosphere under equilibrium conditions and is appropriate only for dilute solutions which are typically observed in the environment. Certain hydrocarbons despite possessing relatively low vapor pressures, may tend to partition significantly toward the air. This is largely a result of their correspondingly low water solubilities which result in low values for K_w. Therefore, chemicals which have low values for K_w have a greater tendency to partition towards the air and volatilize from solution.

Bioconcentration Factor - Fish/Water (BCF). The partitioning of a chemical between water and fish is yet another expression of the hydrophobic nature of the chemical. The ratio of chemical in the fish to that in the water at equilibrium is defined as the bioconcentration factor.

$$BCF = \frac{\mu g \text{ chemical/g fish}}{\mu g \text{ chemical/g water}} \qquad (4)$$

where

BCF = Bioconcentration factor
μg chemical/g fish = concentration of chemical in fish
μg chemical/g water = concentration of chemical in water

The bioconcentration factor, although usually related to fish is actually an estimate of the bioaccumulation potential for biota in general. Different organisms may bioconcentrate a given chemical to a lesser or greater degree, however with different chemicals, the relative ranking with respect to bioconcentration will be essentially the same for all species.

Correlations Between Partition Coefficients. As has been previously discussed, environmental partition coefficients are to a large extent a measure of a chemical's tendency to partition between aqueous and organic media.

Correlations between various combinations of partition coefficients have appeared in the literature. Water solubility has

been related to n-octanol/water ratios(12, 13, 14), bioconcentration factors(15, 16) and soil sorption constants(17-20). N-octan-
ol/water ratios have been correlated with bioconcentration factors
(21, 22, 23) and soil sorption constants have been correlated with
n-octanol/water ratios(18, 19, 24). More recently Kenega and
Goring have given correlation equations for all combinations of
these parameters(25).

The following correlation equations were used in the estimation of partition coefficients used in this paper.

$$\ln WS(ppm) = -1.7288 \ln K_{oc} - 0.01(MP-25) + 15.1621 \quad (20)$$
$$\ln K_{oc} = \ln K_{ow} - 0.7301 \quad (24)$$
$$\ln BCF = 0.935 \ln K_{ow} - 3.443 \quad (25)$$

The advantages of developing such correlations is that once
any of the parameters is known it is then a simple process to
estimate the others. This is particularly useful in early evaluation of chemical partitioning in the environment. From a limited
amount of information on a chemical, for example, its vapor pressure, water solubility and melting point, other partitioning parameters can be estimated and used in simple ecosystem models to
evaluate the chemical's expected environmental distribution.

Partitioning In Model Ecosystems

An ecosystem can be thought of as a representative segment or
model of the environment in which one is interested. Three such
model ecosystems will be discussed (Figures 1 and 2). A terrestrial model, a model pond, and a model ecosystem, which combines
the first two models, are described in terms of equilibrium
schemes and compartmental parameters. The selection of a particular model will depend on the questions asked regarding the chemical. For example, if one is interested in the partitioning behavior of a soil-applied pesticide the terrestrial model would be
employed. The model pond would be selected for aquatic partitioning questions and the model ecosystem would be employed if overall
environmental distribution is considered.

Partition coefficients can then be combined to describe the
ecosystem, assuming all the compartments are well mixed such that
equilibrium is achieved between them. This assumption is generally not true of an environmental system since transfer rates
between compartments may be slower than transformation rates within compartments. Therefore, equilibrium is never truly approached, except for perhaps with very stable compounds. However,
such simplifications can give an indication into which compartments a chemical will tend to migrate and can provide a mechanism
for ranking and comparing chemicals.

Consider the model ecosystem in Figure 2, chosen to represent
a slice of the environment. The dimensions have been selected to
represent a 1000 m x 1000 m square surface which contains a 10 km

MODEL ENVIRONMENTS

Terrestrial:

Soil Air $\xrightleftharpoons{K_w}$ Soil Water $\xrightleftharpoons{K_{oc}}$ Soil Organic Carbon

1000m x 1000m x .076m = 7.6 x 10^4 m^3

25% Air, 25% Water, 50% Soil Solids
Soil Organic Carbon = 2%

Pond:

Sediment $\xrightleftharpoons{K_{oc}}$ Water $\xrightleftharpoons{BCF}$ Fish

Water	1000m x 1000m 10m = 1 x 10^7 m^3
Sediment	1000m x 1000m x .05 = 5 x 10^4 m^3
Suspended Sediment	10 ppm in water = 50m^3
Fish	1 ppm in water = 10m^3
Sediment Organic Carbon	= 4%

Figure 1.   Terrestrial and pond models.

ECOSYSTEM:

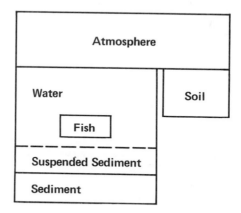

$$\text{Sediment} \xrightleftharpoons[\;]{K_{oc}} \text{Water} \xrightleftharpoons[\;]{K_w} \text{Air} \xrightleftharpoons[\;]{K_w} \text{Soil Water} \xrightleftharpoons[\;]{K_{oc}} \text{Soil}$$

BCF

Fish

Atmosphere	1000m x 1000m x 10Km = $10^{10}\,m^3$
Water	1000m x 300m x 10m = $3 \times 10^6\,m^3$
Soil	1000m x 700m x .076m = $5.4 \times 10^4\,m^3$
Sediment	1000m x 300m x .05m = $1.5 \times 10^4\,m^3$
Suspended Sediment	10 ppm in water = $15m^3$
Fish	1 ppm in water = $3m^3$
Soil Organic Carbon	= 2%
Sediment Organic Carbon	= 4%

Figure 2. Model ecosystem.

column of air. The surface area is assumed to be 30% water, consisting of a pond 10 m in depth. The top 7.5 cm of soil and the top 5 cm of sediment are assumed to represent that portion of those compartments which take part in the equilibrium process.

The overall equilibrium expression for the system can be represented as follows:

$$C_{sed} \xrightleftharpoons{Kd_{sed}} C_w \xrightleftharpoons{K_w} C_a \xrightleftharpoons{K_w} C_{sw} \xrightleftharpoons{Kd_{soil}} C_s \quad (5)$$

$$\updownarrow BCF$$

$$C_f$$

where

C_{sed} = concentration of chemical in sediment
C_w = concentration of chemical in water
C_f = concentration of chemical in fish
C_a = concentration of chemical in air
C_{sw} = concentration of chemical in soil water
C_s = concentration of chemical in soil

The primary compartment that connects the aquatic and terrestrial segments of the ecosystem is the air. The air is considered to be in equilibrium with the soil water (which is assumed to be 25% of the soil compartment) and the water in the aquatic segment.

If a given volume percent for each compartment is assumed, the partition expressions must be written in terms of volumes by considering densities (p) of the media. Soil and sediment are assumed to have particle density of 2.5 g/cc. Water and fish are assumed to have density of 1 g/cc. Air is already expressed on volume basis.

If the amount of chemical in each compartment is expressed as a percentage of the total chemical in the system (M), and if the volume of each compartment is also expressed as a percentage of the total volume of the system (V), then the equilibrium constant between compartments can be written as follows.

$$Kd_{sed} = \frac{\%M_{sed}/(\%V_{sed})\,2.5}{\%M_w/\%V_w} \quad (6)$$

$$BCF = \frac{\%M_f/\%V_f}{\%M_w/\%V_w} \quad (7)$$

$$K_w = \frac{\%M_w/\%V_w}{\%M_a/\%V_a} = \frac{\%M_{sw}/\%V_{sw}}{\%M_a/\%V_a} \quad (8)$$

$$K_{ds} = \frac{\%M_s/(\%V_s)\,2.5}{\%M_{sw}/\%V_{sw}}$$

These expressions can be combined to show that . . .

$$\%M_w = \frac{100}{Kd_{sed}\left(\frac{\%V_{sed}}{\%V_w}\right)2.5 + BCF\left(\frac{\%V_f}{\%V_w}\right) + 1/K_w\left(\frac{\%V_a}{\%V_w}\right) + \left(\frac{\%V_{sw}}{\%V_w}\right) + Kd_s\left(\frac{\%V_s}{\%V_w}\right)2.5} \quad (10)$$

Therefore, the percent of chemical in the water can be solved for, and having this value the percent of chemical in each compartment can be subsequently calculated. Equations for the terrestrial model and the model pond are solved similarly. Also, the sizes of the compartments can be changed to represent different types of environmental conditions.

Further insight into the distribution values obtained may be gained by transforming the percent of the total chemical in each compartment into concentrations. This is done by specifying a load of chemical into the system and calculating concentrations based on the amount of chemical in each compartment and the volumes of the compartments.

Applications of The Model

Seven chemicals, whose physical properties and partition coefficients are shown in Table I were evaluated with all three models (Tables II-IV). When measured values were not available, values were estimated from correlation equations previously described. Inspection of the results reveals several interesting aspects of this type of approach. The compounds are arranged in the order of increasing water solubility. correspondingly, as previously discussed, values for the soil sorption constant (K_{oc}) and the bioconcentration factor (BCF) generally decrease. In the terrestrial model and model pon' the amount of chemical in the water compartment is therefore generally observed to increase. In the terrestrial system the amount of chemical in this compartment can serve as an index to relate the relative leachability of chemicals. In the model pond, chemicals which partition more into the water will in general be less persistent in this type of environment since they will be subjected to a greater degree to dissipation forces of degradation, volatility, dilution, etc. Bioconcentration into the fish also generally decreases with increase in the amount of chemical in the water compartment as a result of a decrease in chemical hydrophobicity. Tetrachlorobiphenyl represents an exception to this trend primarily as a result of an unusually high BCF relative to its water solubility and other properties.

The water to air ratio (K_w) when taken by itself can be used to represent the volatility of chemicals from aqueous solution. As previously discussed according to Henry's Law, as water solubility decreases and vapor pressure increases volatility will increase. Therefore a chemical like tetrachlorobiphenyl will have a

Table I. Physical properties and partition coefficients.

Chemical	MW	WS(ppm)	Vp(mmHg)	K_W	K_{OC}	BCF
DDT	354.5	0.0017	1.9×10^{-7}	470	150,000	61,600
Tetrachlorobiphenyl	290	0.017	4.9×10^{-4}	2	32,500*	72,950
Lindane	290.8	0.15	3.2×10^{-5}	300	1,300	325
Chlorpyrifos	351	1.2	1.9×10^{-5}	3340	6,100	470
Nitrapyrin	230.9	40	2.8×10^{-5}	1150	560	82 [a]
2,4-D	221	900	6.0×10^{-7}	1.3×10^{8}	60	3 [a]
1,3-Dichloropropene	111	2700	25	18	68*	3 [a]

[a] Estimated properties.

Table II. Chemical distribution using terrestrial model.

Chemical	DDT	Tetrachloro-biphenyl	Lindane	Chlorpyrifos	Nitrapyrin	2,4-D	1,3-Dichloro-propene
% in air	1.4×10^{-5}	1.4×10^{-2}	2.6×10^{-3}	4.9×10^{-5}	1.5×10^{-3}	1.1×10^{-7}	1.51
% in water	6.7×10^{-3}	0.031	0.76	0.164	1.75	14.3	27.4
% in soil	99.9	99.9	99.2	99.8	98.2	85.7	71.1
Air Conc. $\mu g/m^3$	0.015	14.6	2.68	0.051	1.61	1.2×10^{-4}	1590
Water Conc. (ppm)	0.0007	0.003	0.080	0.017	0.18	1.50	2.88
Soil Conc. (ppm)	2.11	2.10	2.09	2.10	2.07	1.80	1.50

Table III. Chemical distribution using model pond.

Chemical	DDT	Tetrachloro-biphenyl	Lindane	Chlorpyrifos	Nitrapyrin	2,4-D	1,3-Dichloro-propene
% in water	1.31	5.77	60.6	24.7	78.1	96.8	96.7
% in sed.	98.6	93.8	39.4	75.3	21.9	3.16	3.22
% in su. sed.	0.099	0.094	0.039	0.075	0.022	0.003	0.003
% in fish	0.081	0.42	0.020	0.011	0.0064	3.05×10^{-4}	3.15×10^{-4}
Water Conc.(ppm)	2.63×10^{-4}	1.15×10^{-3}	1.21×10^{-2}	4.94×10^{-3}	1.56×10^{-2}	1.92×10^{-2}	1.93×10^{-2}
Sed. Conc.(ppm)	1.58	1.50	0.63	1.20	0.35	0.051	0.052
Su. Sed. Conc. (ppm water)	1.97×10^{-5}	1.88×10^{-5}	7.88×10^{-6}	1.51×10^{-5}	4.37×10^{-6}	6.32×10^{-7}	6.55×10^{-7}
Fish Conc.(ppm)	16.2	84.2	3.94	2.32	1.28	0.061	0.063

Table IV. Chemical distribution using model ecosystem.

Chemical	DDT	Tetrachloro-biphenyl	Lindane	Chlorpyrifos	Nitrapyrin	2,4-D	1,3-Dichloro-propene
% in air	4.72	97.9	83.2	12.8	65.4	0.0025	99.4
% in water	0.65	0.065	7.48	12.8	22.5	93.8	0.54
% in soil	44.8	0.96	4.41	35.3	5.78	3.14	0.02
% in sed	49.8	1.06	4.86	39.1	6.31	3.06	0.02
% in su. sed.	0.05	0.001	0.004	0.04	0.006	0.003	1.8×10^{-5}
% in fish	0.04	0.0048	0.0024	0.0060	0.0078	0.0003	1.7×10^{-6}
Air Conc. (μg/m^3)	9.43×10^{-3}	1.96×10^{-1}	1.66×10^{-1}	2.56×10^{-2}	1.31×10^{-1}	4.96×10^{-6}	1.99×10^{-1}
Water Conc. (ppm)	4.42×10^{-4}	4.35×10^{-5}	4.99×10^{-3}	8.55×10^{-3}	1.50×10^{-2}	6.25×10^{-2}	3.60×10^{-4}
Soil Conc. (ppm)	1.16	0.023	0.11	0.87	0.14	0.078	4.7×10^{-4}
Sed. Conc. (ppm)	2.65	0.057	0.26	2.09	0.34	0.16	9.7×10^{-4}
Su. Sed. Conc. (ppm water)	3.32×10^{-5}	7.1×10^{-8}	3.2×10^{-6}	2.61×10^{-5}	4.21×10^{-6}	2.04×10^{-6}	1.2×10^{-8}
Fish Conc. (ppm)	27.3	3.18	1.62	4.02	1.23	0.20	0.001

high potential to volatilize from aqueous solution, in fact, somewhat higher than 1,3-dichloropropene which has a vapor pressure
nearly five orders of magnitude greater. In considering volatility in the terrestrial model, K_w is modified by K_{oc} such that the
amount of chemical calculated to be in the soil air can represent
a relative volatility index. In this case, 1,3-D will be much
more volatile than the others. This chemical is used as a soil
fumigant and migrates easily through the soil as a vapor. In
general, chemicals with concentrations greater than 1 $\mu g/m^3$ in
this system with a specified load of 200 kg will tend to be somewhat volatile, therefore nitrapyrin, lindane and tetrachlorobiphenyl will exhibit some degree of volatility.
      Considering the terrestrial model or model pond alone can
give some insight into expected behavior in these systems. However, only by combining all the compartments in an overall ecosystem can a general overview of expected environment behavior be
obtained. For example, the model pond indicates the high potential of tetrachlorobiphenyl to bioconcentrate. However, in the
model ecosystem this chemical tends to migrate toward the atmospheric compartment such that its tendency to bioaccumulate is
greatly diminished. Likewise, 1,3-dichloropropene partitions
almost completely into the air despite its very high water solubility. DDT, on the other hand, partitions mainly into the soil
and the sediment. As a result, concentration of chemical in fish
is relatively high since they share the same aquatic environment
with the sediment.

Advancement of Equilibrium Models

      Discussion to this point has presented equilibrium modeling
in its simplest form or level I as it is termed by Mackay(2).
From this fundamental level the model can be advanced to more complex levels. Inclusion of the dynamics of flow or transfer rates
between compartments and degradation properties within compartments can transform the model to a nonequilibrium, steady state
description of a chemical's fate.
      The next level of complexity is to maintain the assumptions
of the fundamental model, that compartments are well mixed and
rapidly equilibrated, and consider degradation rates within compartments. If this is done, the half-life of the chemical in the
system can be estimated along with an estimated amount degraded in
each compartment.
      First order rate constants are assumed for all degradative
processes: soil and water microbial degradation, hydrolysis, oxidation, photodegradation in air and water and any other mechanisms
of transformation that may apply. The rate at which the chemical
degrades will then be equal to the summation of the rate constants
acting on the amount of chemical in each compartment summed over
all compartments.

The following expression calculates the half-life of the chemical in the model ecosystem.

$$t_{\frac{1}{2}} = 0.693 / \frac{\Sigma k_s (\%M_s) + \Sigma k_{sed}(\%M_{sed}) + \Sigma k_w(\%M_w) + \Sigma k_a(\%M_a)}{\%M_s + \%M_{sed} + \%M_w + \%M_a} \quad (11)$$

where

Σk_i represents the sum of the rate constants acting in the ith compartment.

and $\%M_i$ = the percent of the total chemical in the ith compartment.

An example of the evaluation of a theoretical chemical is shown in Figure 3.

In addition to dissipation of the substance from the model system through degradation, other dissipative mechanisms can be considered. Neely and Mackay([26]) and Mackay([3]) have also introduced advection (loss of the chemical from the troposphere via diffusion) and sedimentation (loss of the chemical from dynamic regions of the system by movement deep into sedimentation layers). Both of these mechanisms are then assumed to act in the unit world. This approach makes it possible to investigate the behavior of atmosphere emissions where advection can be a significant process. Therefore, from a regulatory standpoint if the emission rate exceeds the advection rate and degradation processes in a system, accumulation of material could be expected. Based on such an analysis reduction of emissions would be called for.

The model can be developed further by introduction of transfer rates between compartments. The system no longer is considered to be at equilibrium and diffusion processes become important. Ideally this is the goal of the advanced model, to totally describe the movement of the chemical through and between compartments and into and out of the system as well as describing degradation processes that act in each compartment. This is the greatest level of complexity. Unfortunately, not enough information is available to accurately define the transfer rates. Understanding of volatilization of chemicals from soil needs to be developed further. Transfer rates from water to air can be described in laboratory studies but effects of wind and wave action, mixing and volume relationships needs to be examined furthur. In addition, little is known regarding rates of sorption and desorption of chemicals to soils and sediments.

Finally, degradation processes which are usually assumed to be first order are not. Degradation in soil has been shown by Hamaker ([27]) to often behave in a biphasic manner. Biodegradation in water has been shown to more closely follow second order kinetics([28]). Photolysis in solution is highly dependent on antenuation of light in the water body which will depend on water quality

Chemical: X

Physical Properties

* Molecular Weight= 300
  Water solubility= 1.23E+01
* Vapor pressure = 1.00E-05
* Melting Point = 100

Partition Coefficients

* Koc= 1000
  Kw = 76337
  Kow= 2075
  BCF= 40

PERCENT OF CHEMICAL IN EACH COMPARTMENT

	Air (1.00E+10	Water 3.00E+06	Sediment 1.50E+04	Sp. Sediment 1.50E+01	Fish 3.00E+00	Soil 5.40E+04)
Percent	2.19E+00	5.00E+01	2.50E+01	2.50E-02	2.02E-03	2.27E+01

CONCENTRATION OF CHEMICAL IN EACH COMPARTMENT IN ppm

Air (micro-g/cu-mtr)	Water	Sediment	Sp. Sediment	Fish	Soil
4.37E-03	3.34E-02	1.33E+00	1.67E-05	1.35E+00	5.62E-01

COMPARTMENT HALF-LIVES

Soil	Water	Sediment	Air
300	200	150	50

CHEMICAL HALF-LIFE IN TOTAL SYSTEM = 186 DAYS

* = Entered values, other values are estimated

Figure 3. Estimation of physical properties and partition coefficients of organic chemicals.

and depth(29) and photolysis in air is often catalyzed by free
radical reactions(26).

In summary, a great deal of reasearch is yet to be conducted
to describe these complex phenomena.

Advantages and Disadvantages of Equilibrium Models

At the fundamental level of equilibrium modeling the advantages are many. The model can combine a number of compartments
through simple relationship to describe a realistic environment
within which chemicals can be ranked and compared. Primary compartments that chemicals will tend to migrate toward or accumulate in can be identified. The arrangement of compartments and
their volumes can be selected to address specific environmental
scenarios. Data requirements are minimal, if the water solubility and vapor pressure of a chemical are known, other properties
can be estimated, and a reasonable estimate of partitioning characteristics can be made. This is an invaluable tool in the early
evaluation of chemical, whether the model be applied to projected
environmental hazard or evaluation of the behavior of a chemical
in an environmental application, as with pesticides. Finally,
the approach is mathematically very simple and can be handled on
simple computing devices.

In addition, this first simple stage can act as a starting
point for more advanced models which consider transfer and transformation processes in a more comprehensive manner. It provides
a focal point for sorting out key information regarding a chemicals fate in the environment, and acts as good first approximation of behavior upon which to base further research.

Some disadvantages have already been mentioned. These primarily appear as the model is made more complex. When degradation processes are considered at the next highest level (level
II) care must be taken with interpretation of the data, in particular with less persistent compounds. 2,4-D for example, when
applied to soil or a terrestrial system degrades very rapidly,
much more rapidly than in water. If the half-life of the chemical
was evaluated in the model ecosystem, it would be overestimated
since the majority of the chemical tends to equilibrate in the
water compartment. Relatively stable compounds for which transfer
rates will be faster than dissipative rates can be evaluated more
realistically.

A second shortcoming that arises at this stage of evaluation
is that in order to conduct the evaluation much more information
is required, i.e. soil and sediment degradation rates and hydrolysis and photolysis rates. At this point, more complex nonequilibrium models may be more useful. If and when methods of estimating degradation process become available, this level of evaluation will become more useful.

The disadvantages of the nonequilibrium steady state models
have already been pointed out. In addition, evaluative models of

this type become more and more specific as more information is
required to define the system and in so doing some usefulness may
be lost.

Summary

     Equilibrium partitioning models can be a valuable tool in the
early evaluation of a chemical's fate. In an atmosphere of continued regulatory demands and a search for knowledge and understanding of chemical behavior in the environment, equilibrium
models provide a meaningful starting point for the evaluation process. They do not attempt to supply a description of the ultimate
fate of a compound, but do lead to advanced understanding and direction in sorting out the information required to meet this goal.

Literature Cited

1. Hamaker, J. W. "Diffusion and Volatilization"; in Organic
    Chemicals in The Soil Environment, C. A. I. Goring and J. W.
    Hamaker Eds. Marcel Dekker, Inc., New York. 1972.
2. Mackay, D. Environ. Sci. Technol. 1979, 13, 1218-1223.
3. Mackay, D.; Paterson, S. Environ. Sci. Technol. 1981, 15,
    1006-1013.
4. Baughman, G. L.; Lassiter, R. ASTM STP 657, Philadelphia.
    1978, 35.
5. McCall, P. J.; Laskowski, D. A.; Swann, R. L.; Dishburger, H.
    J. Residue Reviews in press.
6. Lindstrom, F. T.; Hague, R.; Freed, V. H.; Boersma, L. Envi-
    ron. Sci. Technol. 1967, 1, No. 7, 561-565.
7. Kay. B. D.; Elrick, D. E. Soil Science 1967, 104, 314-322.
8. Oddson, J. K.; Letey, J.; Weeks, L. V.; Soil Sci. Soc. Amer.
    Proc. 1970, 34, 412-417.
9. Lindstrom, F. T.; Boersma, L.; Stockard, D. Soil Science
    1971, 112, 291-300.
10. Huggenberger, F.; Lety, J.; Farmer, W. J. Soil Science Soc.
    Amer. Proc. 1974, 36, 544-548.
11. Dilling, W. L. Environ. Sci. Technol. 1977, 11, 405-409.
12. Chiou, C. T.; Freed, V. H.; Schmedding, D. W.; Kohnert, R. L.
    Environ. Sci. Technol. 1977, 11, 475-479.
13. Yalkowski, S. H.; Valvani, S. C. J. Chem. Eng. Data 1979,
    24, 108.
14. Yalkoswki, S. H.; Orr, R. J.; Valvani, S. C. Ind. Eng. Chem.
    Fundam. 1979, 18, 351-353.
15. Fujita, T.; Iwasha, J.; Hansch, C. J. J. Am. Chem. Soc.
    1964, 86, 5175-5180.
16. Leo, A.; Hansch, C. J.; Elkins, D. Chem. Reviews 1971, 71,
    525-616.
17. Chiou, C. T.; Peters, L. J.; Freed, V. H. Science 1979, 206,
    831-832.
18. Karickhoff, S. W.; Brown, D. S.; Scott, T. A. Water Res.
    1979, 13, 241-248.

19. Hassett, J. J.; Means, J. C.; Barnwart, W. L.; Wood, S. G. EPA-600/3-80-041, 1980.
20. Swann, R. L.; Laskowski, D. A.; McCall, P. J.; VanderKuy, K. Second Chemical Congress of the North American Continent. San Francisco, 1980.
21. Neely, W. B.; Branson, D. R.; Blau, G. E. Environ. Sci. Technol. 1974, 8, 1113-1115.
22. Tulp, M. Th. M.; Hutzinger, O. Chemosphere 1978, Univ. Amsterdam, the Netherlands, 12.
23. Hamelink, J. L.; Waybrant, R. C.; Ball, R. C. Trans. Amer. Fish Soc. 1971, 100, 207.
24. Briggs, G. G. Proceedings 7th British Insecticide and Fungicide Conference, 1973, 83-86.
25. Kenaga, E. E.; Goring, C. A. I. Third ASTM Symposium on Aquatic Toxicology, New Orleans: ASTM Philadelphia, 1978.
26. Neely, W. B.; Mackay, D. 1981, presented at workshop "Modeling the Fate of Chemicals in the Aquatic Environment", Pellston, Michigan.
27. Hamaker, J. W.; Colloginm on Terrestrial Microcosoms, ACS Symposium Series, 1977.
28. Paris, D. F.; Steen, W. C.; Baughman, G. L.; Barnett, J. T.; Appl. Environ. MicroBiol. 1981, 41, 603-609.
29. Zepp, R. G.; Cline, D. M. Environ. Sci. Technol. 1977, 11, 359-366.

RECEIVED April 26, 1983.

A New Mathematical Modeling System

R. C. JOHANSON

University of the Pacific, Dept. of Civil Engineering, Stockton, CA 95211

The Hydrological Simulation Program - Fortran (HSPF) simulates the movement of water, sediment and many associated constituents over and under the land surface, and through streams, rivers and shallow lakes. It is a relatively new package, funded by the U.S.EPA. Substances such as pesticides are handled by simulating the processes of adsorption and desorption, transport in the dissolved and adsorbed states, and degradation through hydrolysis, photolysis, oxidation by free radicals, etc. Spatial variation is handled by subdividing the area into "processing units", each of which represents a relatively homogeneous segment of pervious or impervious land-surface, or a reach of channel or an impoundment. HSPF operates on the basis of continuous simulation; the user can select a time step ranging from 1 minute to 1 day. The software package incorporates many modern features which have made it reliable and easy to install on a variety of machines. It is also easy to use and extend.

As environmental controls become more costly to implement and the penalties of judgment errors become more severe, water quality management requires more efficient analytical tools based on greater knowledge of the phenomena to be managed. In this connection, the development and application of mathematical models to simulate the transport and transformation of pollutants through a watershed, and thus to anticipate environmental problems, has been the subject of intensive research by the Environmental Research Laboratory (U.S. EPA) in Athens, Georgia. HSPF is one of the most recent products of this work. Starting in 1976, it was developed from the following older models:
(1) The Stanford Watershed Model (SWM) developed at Stanford University (1). It can simulate the hydrologic behavior of an entire watershed.

0097–6156/83/0225–0125$06.75/0
© 1983 American Chemical Society

(2) The Agricultural Runoff Management (ARM) Model, developed by
     Hydrocomp Inc. for the U.S.EPA (2). It simulates the
     hydrology, sediment yield, and nutrient and pesticide
     behavior of the land phase of the hydrological cycle. The
     same organizations also developed the Non-Point Source (NPS)
     Model (3) which handles the washoff of miscellaneous
     pollutants from land surfaces.
(3) The HSP Quality Model (4). It simulates a comprehensive set
     of water quality processes in streams and lakes, but not
     pesticides and toxic substances.

Later some features of the SERATRA Model, developed by
Batelle Northwest Laboratories (5) were added. This model was
designed to simulate the behavior of sediment and associated
constituents in streams. It includes processes such as
hydrolysis and photolysis and is thus suitable for modeling toxic
substances such as pesticides.

Basic Principles of HSPF

All of the above models are of the "deterministic conceptual"
type. That is:
(1) they do not contain random components. A given set of input
     data will always produce the same set of output.
(2) they consist of sets of linked equations which represent, to
     a certain degree, the actual phenomena being simulated.

These models employ continuous, rather than single event,
simulation. The advantage is that continuous output can be
analyzed statistically. The user can obtain answers to questions
such as "For what fraction of time will the concentration of X be
above Y mg/l at point Z in the system?" Or, "What danger does
chemical X pose to species A at locations B and C?". These are
the kinds of answers needed if he is to make rational decisions
regarding the permissible uses of chemicals for agricultural
purposes.
HSPF represents the temporal variations in a basin by
simulating its behavior over an extended period of time, using a
constant time step selected by the user. Spatial variations are
handled by subdividing the basin into several distinct
computational elements or Processing Units (PUs) (Figure 1).
There are several types of PUs such as:
(1) Pervious Land-segments, simulated by the PERLND module.
(2) Impervious Land-segments, simulated by the IMPLND module.
(3) Free-flowing stream reaches and reservoirs, simulated by the
     RCHRES module.

The degree to which the study area is subdivided is up to
the user; HSPF can handle hundreds of PUs in a single run. For

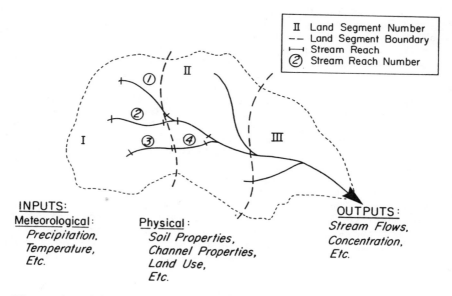

Figure 1. Subdivision of a basin for simulation using HSPF.

every PU the hydrologic response is first simulated. Then calculations of water temperature, sediment transport and chemical behavior are superimposed on the flow calculations. The user specifies how the various PUs are connected, forming the "network" of water, constituent and information flow (Figure 1).

There are two classes of Operating Module in HSPF:

(1) Application Modules. These simulate the behavior of processes which occur in the real world (eg. a Pervious Land-segment).

(2) Utility Modules. These perform "housekeeping" operations on time series (eg. multiply a concentration time series by a flow time series, to get a "load" time series).

HSPF is an expandable system. Operating modules can be added to, or removed from, the system with relative ease. The software currently contains three application modules and six utility modules (Figure 2).

Both classes of Operating Module usually need one or more input time series and produce one or more output time series (eg. outflow of water and constituents). From experience, the designers of HSPF knew that much of the effort in using continuous simulation models is associated with time series manipulations. Thus, a sophisticated Time Series Management System was included. It centers around the Time Series Store (TSS) (Figure 10), which is a disk-based file on which any input or output time series can be stored indefinitely.

HSPF can be run with a time step ranging from 1 minute to 1 day. Data can be stored in the TSS with a similar range of intervals. The system will automatically convert time series from one interval to another, as they are transferred between the TSS and the machine memory. This means, for example, that a Pervious Land-segment could be run at an interval of 1 hour, using 15 minute precipitation data and daily evaporation data (stored on the TSS) as inputs.

The Pervious Land-segment (PERLND) Module

General Comments. The PERLND module simulates a variety of processes occurring on and under the surface of a Pervious Land-segment. Figure 3 is a "structure chart" (see "Software Considerations") which shows the twelve sections of this module and the functions they perform. The sections usually involved in simulating pesticides are SNOW and PWATER (hydrology), SEDMNT (sediment), MSTLAY (solute transport) and PEST (pesticides). The last 5 sections of the module are of primary importance in simulating agricultural chemicals.

The user specifies which set of sections will be executed in a given run. For example, he may initially "switch on" only SNOW and PWATER, to calibrate the simulated hydrological behavior of a land-segment to observed data. Then he may turn on MSTLAY and

Application Modules		
PERLND	IMPLND	RCHRES
Snow	*Snow*	*Hydraulics*
Water	*Water*	*Conservative*
Sediment	*Solids*	*Temperature*
Quality	*Quality*	*Sediment*
Pesticide		*Nonconservative*
Nitrogen		*BOD/DO*
Phosphorus		*Nitrogen*
Tracer		*Phosphorus*
		Carbon
		Plankton

Utility Modules		
COPY	PLTGEN	DISPLY
Data transfer	*Plot data*	*Tabulate, summarize*
DURANL	GENER	MUTSIN
Duration Analysis	*Transform or combine*	*Input sequential Time-series data*

Figure 2. "Operating Modules" presently in the HSPF software.

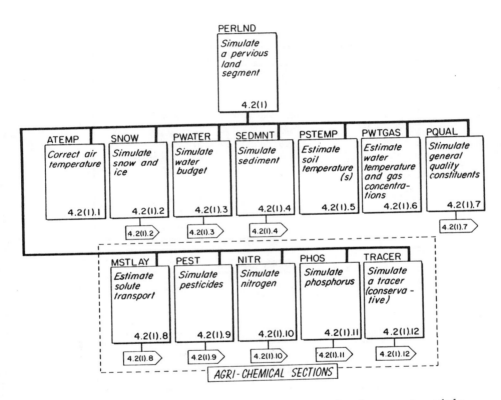

Figure 3.   Structure chart for the Pervious Land–segment module.

TRACER so that he can compare the simulated and observed movement of a conservative substance such as chloride. Finally, he may turn TRACER off and PEST on, to simulate up to 3 pesticides.

Hydrologic Simulation in the PERLND Module.   Hydrologic simulation is done using the moisture accounting technique first employed in the Stanford Watershed Model (Figure 4). That is, the movement of water into, between, and out of, a set of conceptual storages is computed using a fixed time step. Snow accumulation and melt are simulated in the SNOW section (if it is turned on) using energy balance procedures (6). Rain and snowmelt are subject to interception. If that storage is full infiltration occurs. Infiltration capacity is a function of the current moisture storage in the lower zone and a parameter INFILT which reflects the permeability of the soil. Infiltrated moisture passes to the lower zone or to groundwater storage. Excess moisture either remains on the surface or enters flow paths leading to the upper zone or to interflow. Percolation from the upper zone to the lower zone and groundwater is modeled. The model regards overland flow as equivalent to that along a plane surface of length, slope and roughness specified by the user. It uses a kinematic method to calculate the overland flow rate. Other contributions to streamflow come from interflow and groundwater outflow.

Evapotranspiration (ET) can occur from any of the storages. The model algorithms compute the amount of ET from each storage, based on potential ET data supplied by the user.

Sediment Simulation in the PERLND Module. The processes modeled in the SEDMNT section are shown in Figure 5. It also shows the simple equations used, which are based on one of the first continuous sediment simulation models (7). The rate of detachment by rainfall is a power function of rainfall intensity, modified to account for protective cover (C) and any special management practices (SMPF) (e.g. terracing, contouring). SMPF corresponds to the factor P in the Universal Soil Loss Equation. Washoff (WS) is the removal, by overland flow, of detached material. It is modeled as a power function of overland flow, which is computed by the hydrology section (PWATER), but washoff is limited by the supply of detached material. This supply can be altered by the user at any time, to simulate the effect of soil tillage. Scour (SCR) is also modeled as a power function of overland (surface) outflow. This simulates direct erosion by surface outflow, such as gully formation. For scour, the model considers the supply of parent material unlimited. The coefficients and exponents (KRER, JRER, etc) must be determined by experience and/or calibration.

The sediment section also accounts for soil compaction (using a first-order process) and deposition or removal of detached sediment (e.g. by wind).

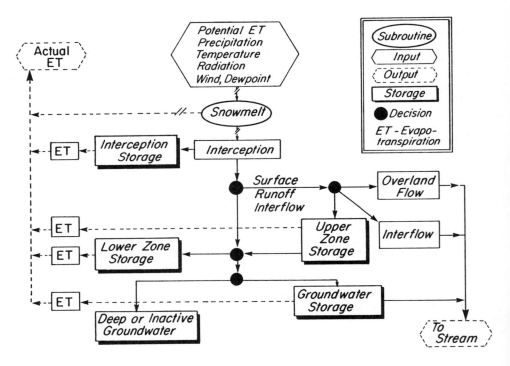

Figure 4. Representation of the hydrological processes in a Pervious Land-segment.

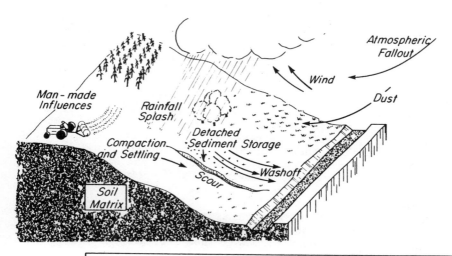

Detachment:	$DET = (1-C) * SMPF * KRER * RAIN^{JRER}$
Washoff:	$WS = KSER * SURO^{JSER}$
Scour:	$SCR = KGER * SURO^{JGER}$

Figure 5. Sediment—related processes in a Pervious Land-segment, as modeled in HSPF.

Pesticide Simulation in the PERLND Module. The procedures used
to simulate pesticides were derived from those used in the
Agricultural Runoff Management (ARM) Model (2).   HSPF can
simulate up to 3 pesticides in one run. The soil is viewed as
having four layers (Figure 6), corresponding to the surface,
upper, lower and groundwater storages used in the hydrology
section (Figure 4). Although in nature the transport and
reactions of pesticides occur simultaneously, the model treats
these processes serially, in each time step.
     Transport rates for dissolved material are based on the
internal and external fluxes (flows) computed in the hydrology
section of the module. Soluble chemicals are transported down
through the soil profile and are washed out into streams with
surface runoff, interflow and groundwater flow. Sediment
associated pesticides (and nutrients) are removed from the
surface layer whenever sediment washoff occurs.
     The two pesticide reactions simulated by HSPF are:
(1) Adsorption and desorption. The user can choose to handle
     this using either temperature-corrected first order reaction
     kinetics, in which case the concentrations are always moving
     towards equilibrium but never quite reach it, or he can use
     a Freundlich isotherm, in which instantaneous equilibrium is
     assumed. With the Freundlich method, he can elect either to
     use a single-valued isotherm or a non-single-valued one.
     This was included in the model because there is experimental
     evidence which suggests that pesticides do not always follow
     the same curve on desorption as they do on adsorption.
(2) Degradation. Although the actual mechanisms of degradation
     are many and complex, HSPF uses a simple first-order
     relationship to approximate this process.

     Adsorption, desorption and degradation are simulated in each
of the four soil layers (Figure 6). Different parameters can be
used in each layer.
     Note that this section of the PERLND module could be used to
simulate substances other than pesticides, provided the processes
included in the model adequately represent the behavior of the
compound in question.

The Impervious Land-segment (IMPLND) Module

This module is designed to simulate processes in areas where the
ground is totally impervious; usually it is used on parts of
urban areas. It is not designed to handle pesticides.

The Reach/reservoir (RCHRES) Module

General Comments. As the structure chart for this module shows
(Figure 7), it is designed to simulate the transport and

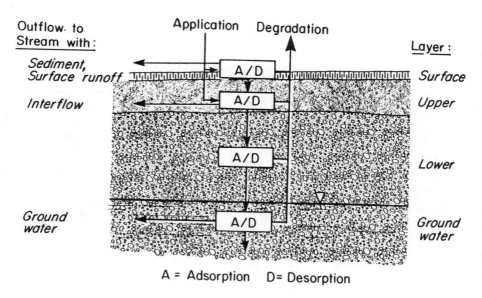

Figure 6.   Pesticide related processes in a Pervious Land-segment, as modeled in HSPF.

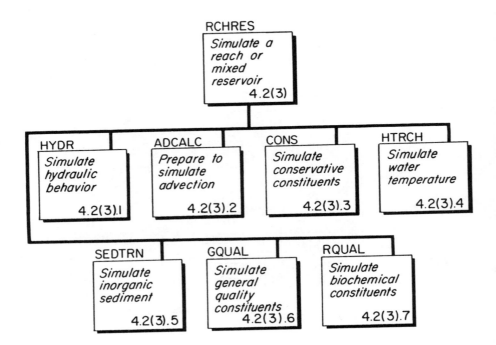

Figure 7.   Structure chart for the Reach/reservoir module.

reactions of a wide variety of constituents in streams and lakes.
Like the PERLND module, each section of the RCHRES module
simulates a different set of processes, and the user can switch
on that combination of sections which is best suited to simulate
the constituents which he is studying.

Section HYDR simulates the movement of water (hydraulic
routing). Section HTRCH evaluates the exchange of heat between a
reach and the atmosphere and, thus, simulates water temperature.
These sections are important because transport and temperature
greatly influence almost all the other processes simulated by the
module. Sections SEDTRN and GQUAL simulate the movement of
sediment and "generalized" quality constituents (e.g.
pesticides). Section RQUAL simulates the "traditional"
biochemical constituents, such as oxygen, biochemical oxygen
demand, nutrients, phytoplankton, zooplankton, carbon dioxide and
refractory organic products of biological death and respiration.

One significant limitation of the RCHRES module is that it
assumes total mixing in the water body; thus it does not simulate
stratified impoundments.

Hydraulic Routing in the RCHRES Module. HSPF uses a simple
technique for flow routing. The catchment stream network is
subdivided into "reaches" (Figure 1) and calculations start with
the upstream ones. Each reach may have several outflows and each
outflow rate may be a function of storage in the reach (storage
routing) or a function of time (e.g. to supply demands of
irrigators), or a combination of both.

HSPF can handle a reach network of any complexity; it can
even handle situations where flows are split (diverted) and later
recombined further downstream (e.g. through hydro-power
diversion tunnels). Also, it makes no assumptions regarding the
shape of the water body. For example, streams do not have to be
represented with trapezoidal cross sections.

Sediment Routing in the RCHRES Module. The sediment routing
method has been adapted from that used in the SERATRA model (5).
Each reach is viewed as containing one "layer" of suspended, or
entrained, sediment and one layer of bed sediment. Three classes
of sediment are handled – sand, silt and clay. Each is
separately routed through the reach and its deposition or erosion
rate is calculated.

For sand, the transport capacity is first calculated using
either the Colby (9), or Toffaleti (10) method, or a user
supplied power function of velocity. If the calculated transport
capacity exceeds the load present scour is simulated and if the
opposite is true deposition is simulated.

For silt and clay, the critical shear stress concept is
used. If the critical shear stress for scour is exceeded, scour
takes place. On the other hand, if the actual shear stress is
less than the critical value for deposition, deposition occurs.

Pesticide Simulation in the RCHRES Module.   Pesticides and many
other toxic substances are subject to a variety of processes in
the aquatic environment.   In the RCHRES module, such compounds
are called "generalized quality constituents" and they are
simulated using module section GQUAL (Figure 7).

The algorithms used by module section GQUAL are, again,
based on those incorporated in the SERATRA model.   The chemical
forms which it can handle and the processes included are shown
schematically in Figure 8.   In this section of the module it is
assumed that all chemicals exist in solution and are, thus,
potentially subject to the processes shown on the left side of
the figure.   These include:
(1) movement with the water (advection).
(2) hydrolysis. A first order pH-dependent equation is used.
(3) oxidation by agents such as singlet oxygen and alkylperoxy
     radicals. A second order equation is used.
(4) volatilization.   This is linked to the oxygen reaeration
     rate which can be computed using a variety of equations.
(5) biodegradation. A second order equation is used.
(6) "other" methods of decay. A first order equation is used.
(7) formation of "daughter" products by decay of "parent"
     compounds.

The user decides which of the above processes will be
simulated (active) for each chemical. He need only supply input
for those processes that are active.   In this connection, note
that:
(1) all of the above decay rates can be adjusted for
     temperature.
(2) much of the supplementary input required for these processes
     (e.g. biomass concentrations, free radical concentration)
     can be supplied either as time series, or as monthly cyclic
     data, or single fixed values.

If the user specifies that the chemical is sediment
associated, then all the processes shown on the right of Figure 8
also become active:
(1) Adsorption and desorption between the solution phase and
     sand, silt and clay in suspension and on the bed.   First
     order reaction kinetics are used.
(2) Transport of adsorbed material with the sediment.   This
     includes advection, scour and deposition.
(3) Decay of adsorbed chemical, modeled as a first-order
     process.

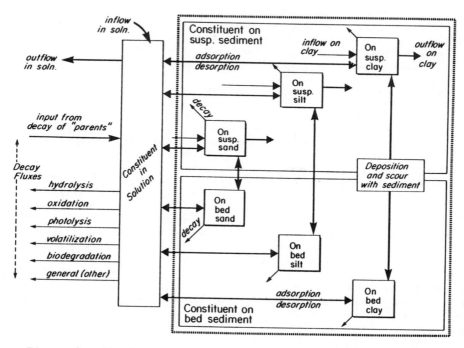

Figure 8. Simulation of a "generalized quality constituent" in HSPF.

The Utility Modules

HSPF's utility modules (Figure 2) are designed to give the user
maximum flexibility in managing simulation input and output.
COPY is used to manipulate time series, such as the transfer of
data from tape to the TSS. The user can change the form of the
time series during the COPY operation. A 5-minute rainfall
record may be aggregated to an hourly time interval, for example.
The PLTGEN and DISPLY modules are discussed later, under
"Output". The GENER module is used to transform a time series
(A) to produce a new series (C) or to combine two time series
(A&B) to create a new one (C). For example, this module is
useful if one wants to compute the mass outflow of a constituent
from the two time series of flow and concentration.

DURANL performs a duration and excursion analysis on a time
series and also computes some statistics. It can answer
questions like: "How often does the DO concentration stay below
4 mg/l for 4 consecutive hours?" This module includes a feature,
derived from Onishi et al (11), for assessing the effect of the
likely exposure of a specified species to a given chemical. It
is presumed that the organism will suffer no damage if the
chemical is always present at levels below the "maximum
acceptable toxicant concentration" (MATC). But if this level is
exceeded the organism will suffer either acute or chronic damage,
depending on the concentration and the time for which it persists
(Figure 9). For example, the borderline for 7-day continuous
exposure might be 1 ppm and the corresponding 1-day value might
be 10 ppm. To perform this type of analysis using HSPF, the user
supplies module DURANL with the data necessary to compose Figure
9, and the time series of chemical concentration values. Then,
in addition to performing the usual statistical analyses, DURANL
determines the percentage of time that acute, chronic and
sub-lethal conditions would exist. The time series can be either
simulated or observed data.

Software Design Considerations

Development of HSPF involved the merging of most of the
capabilities of a set of existing models. This kind of project
can be accomplished in at least two ways. On the one hand, the
existing software can be left largely intact; modules can be
merged using interfaces which require a minimum of new code and
alterations to existing programs. On the other hand, the
functions performed by the older models can be incorporated in a
totally new package. The first approach involves a lower
investment but the shortcomings of the existing models, and any
inconsistencies between them, remain. The second approach is
costly but overcomes these problems; a design can be adopted
which draws on the experience gained in working with the older
models and also incorporates modern program design technology.
For intermediate values of shear stress, the bed is stable.

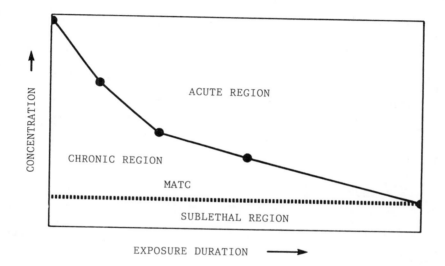

Figure 9.   Lethality analysis of chemical concentration data.

In this way a system can be built which is internally consistent, reliable and relatively easy to use, maintain and expand. The designers (and EPA) chose the second route.

Developing, modifying, or even trying to understand, a large computer program can be a very frustrating activity. To a large extent, the problems can be alleviated by using Structured Programming Technology. Because of its obvious suitability, extensive use was made of this technology on this project. The entire set of software was arranged in hierarchical order, shown on a "structure chart" (eg. Figures 3 and 7). The general idea is that the entire system should form a tree, branching out from the MAIN sub-program. "Continuation flags" point to subordinate structure charts, so that the entire HSPF program can be viewed by studying the 80 structure charts needed to completely describe it. These charts totally supplant the traditional "flowcharts". Within each sub-program, instructions were first coded in "pseudo code," similar to Algol. In accordance with the tenets of structured programming, as developed by Dijkstra and others during the 1960's, the pseudo code included only the following five basic "structure figures:" SEQUENCE, IF-THEN-ELSE, DO-UNTIL, WHILE-DO and CASE. The pseudo code was then translated to standard Fortran. The benefits of writing in structured pseudo code, compared to Fortran, became very obvious as the project progressed.

The entire HSPF system is documented in a User's Manual (12).

Operation of the Model

Overview. Figure 10 shows, in simplified form, the activities, inputs and outputs involved in running HSPF, from a typical user's point of view. The first phase involves copying input time series, such as meteorological data, from sequential files (cards, tape, disc) to the Time Series Store (TSS). This is sometimes done in a single run but in most practical situations, where data have to be gathered from diverse sources and gaps have to be patched, several runs are made.

In the second phase the input time series for simulation, data display and analysis runs usually come from the TSS, although they can sometimes be obtained directly from sequential files, thus bypassing the first phase described above. The other type of input, required in all HSPF runs, is called the User's Control Input.

The User's Control Input. The HSPF system has been made as "intelligent" as possible. For example:
(1) It checks that user-supplied values fall within a reasonable range, where possible.
(2) If a user omits some input, HSPF will supply default values if they exist, or report an error if they do not.

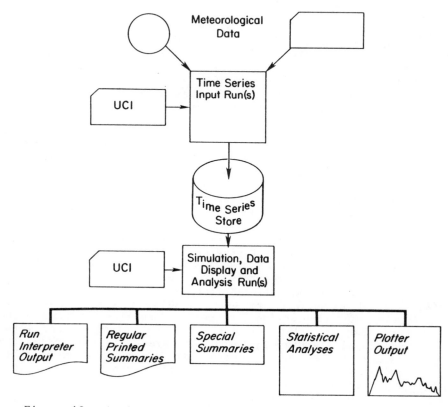

Figure 10.   Activities involved in running HSPF, from the user's viewpoint.

(3)   It will ignore unnecessary input and blank or "comment" lines. Thus, if a user has been simulating pesticides and then turns that section off, possibly to re-calibrate the hydrology, he does not have to delete the pesticide-related input. HSPF will ignore it, until he once again turns the PEST section on.

(4)   It can accept input in Metric or English units. (e.g. pesticide application in kg/ha or lb/acre, rainfall and runoff in mm or inches).

Output. HSPF produces several classes of output:
(1)   Continuous time series. These data are either passed as input to operations further "downstream" in the network or they are recorded on disk in the Time Series Store, or both things may be done (Figure 10).

(2)   Run Interpreter output. This is produced as the User's Control Input is scanned and checked, defaults are supplied, etc. It is, roughly, an echo of the input, plus default values supplied by HSPF.

(3)   Regular printed summaries. Once the simulation time loop is commenced, data are accumulated for display at an interval specified by the user. The frequency of this output can be varied from once per time step (say 1 hour) to once per year. Regardless of the reporting period, the format of the report is the same. First, the values of all significant state variables (eg. storages), at the end of the reporting period, are given. Then, the fluxes (eg. flows), accumulated since the preceding report, are summarized. The user can specify whether printout is to be given in English or Metric units (regardless of the units used for input). Or he may request output in both systems.

(4)   Special summaries. By using module DISPLY, the user may select any time series for special display. For example, he may wish to print out daily average values of the total amount of Alachlor in the Upper Layer. In this case, the values (simulated or observed) would automatically be averaged over each day and a year's worth of daily values would appear in a neatly formatted table on a single page (suitable for direct inclusion in a report).

(5)   Statistical analyses. Again, any time series can be analyzed, by pointing it to module DURANL. The functions of this module have already been described in some detail.

(6)   Plot data. The PLTGEN module can be used to route a set of time series to a file, so that they can later be displayed in graphical form, either together or on separate graphs. In this connection, Figures 11 and 12 show typical pesticide simulation results for a small watershed in Iowa, as reported by Donigian et al (13).

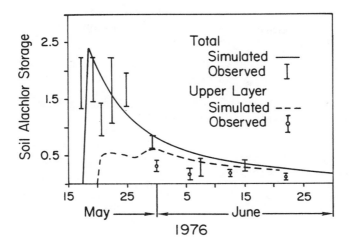

Figure 11.   Verification of simulated soil storage of alachlor in Four Mile Creek watershed, Iowa. (Reproduced with permission from Ref. 13.)

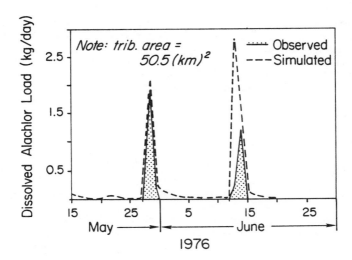

Figure 12.   Verification of simulated flow of alachlor in Four Mile Creek at Traer, Iowa. (Reproduced with permission from Ref. 13.)

Concluding Comments

Is HSPF living up to the designers' expectations? The author's
answer is in the affirmative, for the following reasons:
(1) As predicted, programming errors have been quite rare and
     easy to locate and fix.
(2) Users who have studied the code have found it easy to follow
     – I believe this is an important test of program clarity and
     consistency.
(3) Operating modules and individual subprograms have been added
     or modified with relative ease.
(4) Adaptation of HSPF to computers with widely differing memory
     configurations and word lengths has been relatively simple.
(5) The program has been used in a variety of situations ranging
     from simple single land–segment simulations to studies
     involving complex networks. The constituents handled have
     ranged from water (only) to pesticides, nutrients and
     certain biota. This demonstrates the versatility of the
     program.

In summary, HSPF is an advanced software framework which can
accommodate a wide variety of time series–based simulation
modules and associated utility functions. As better algorithms
for simulating the various natural processes become available,
they can be incorporated into the system. Plans are already
under way to include in the PERLND module an improved technique
for solving the equations governing the advection and reactions
of pesticides and nutrients. The construction of an interactive
input preparer, currently under way, is being greatly facilitated
by the highly structured design of the User's Control Input.
This new software will make the package even easier to use. HSPF
can grow with the state of the art and should have a long and
useful life. It is being maintained, distributed and actively
supported by the EPA Center for Water Quality Modeling, in
Athens, Georgia.

Literature Cited

1. Crawford, N.H., and R.K. Linsley. Digital Simulation in
    Hydrology: Stanford Watershed Model IV. Dept. of Civ. Eng.,
    Stanford Univ., Stanford, Calif. Tech. Rep.39. 1966. 210
    pages.
2. Donigian, A.S., Jr., D.C. Beyerlein, H.H. Davis, Jr., and
    N.H. Crawford. Agricultural Runoff Management (ARM) Model
    Version II: Refinement and Testing. Env. Res. Lab., Athens,
    Georgia. 1977. EPA600/3-77-098. 294 pages.
3. Donigian, A.S. Jr., and N.H. Crawford. Modeling Nonpoint
    Pollution from the Land Surface. Env. Res. Lab., Athens,
    Georgia. 1976. EPA 600/3-76-083. 280 pages.
4. Hydrocomp Inc. Hydrocomp Water Quality Operations Manual.

Hydrocomp Inc., 201 San Antonio Circle, Mountain View, CA 94040. 1977. 192 pages.

5. Onishi, Y,. and S.E. Wise. Mathematical Model, SERATRA, for Sediment-Contaminant Transport in Rivers and its Application to Pesticide Transport in Four Mile and Wolf Creeks in Iowa. Batelle Pacific NW Labs, Richland, WA. 1979.

6. U.S. Army Corps of Engineers. Snow Hydrology, Summary Report of the Snow Investigations. North Pacific Division, Portland, OR. 1956. 437 pages.

7. Negev, M. A Sediment Model on a Digital Computer. Dept. of Civ. Eng., Stanford Univ., Stanford, CA. Tech. Rep.76. 1967. 109 pages.

8. Crawford, N.H. and A.S. Donigian, Jr. Pesticide Transport and Runoff Model for Agricultural Lands. Office of R and D, U.S. EPA, Wash. DC, 1973. EPA 660/2-74-013. 211 pages.

9. Colby, B.R. Practical Computation of Bed-Material Discharge. J. Hyd. Div., ASCE. 1964, 90(HY2),217-246.

10. Toffaleti, F.B. Definitive Computations of Sand Discharge in Rivers. J. Hyd. Div., ASCE. 1969, 95(HY1), 225-248.

11. Onishi, Y., S.M. Brown, A.R. Olsen, M.A. Parkhurst, S.E. Wise and W.H. Walters. Methodology for Overland and Instream Migration and Risk Assessment of Pesticides. Batelle Pacific NW Labs., Richland, WA. Prepared for U.S. EPA, Athens, GA. 1979.

12. Johanson, R.C., J.C. Imhoff and H.H. Davis, Jr. Users Manual for Hydrological Simulation Program - Fortran (HSPF). Env. Res. Lab., Athens, GA. 1980. EPA 600/9-80-015. 678 pages.

13. Donigian, A.S., Jr., J.C. Imhoff and B.R. Bricknell. Modeling Water Quality and the Effects of Best Management Practices in Four Mile Creek, Iowa. Draft report on contract 68-03-2895, for U.S. EPA, Athens, GA. 1981.

RECEIVED April 15, 1983.

MODEL VALIDATION

Model Predictions vs. Field Observations: The Model Validation/Testing Process

ANTHONY S. DONIGIAN, JR.

Anderson-Nichols & Co., Palo Alto, CA 94303

The goal of this paper is to present the current status of <u>model</u> <u>validation</u> and <u>field</u> <u>testing</u> of chemical fate and transport models; other papers in this symposium discuss the state-of-the-art of modeling specific processes, environments, and multimedia problems. The process of model validation, and its various components, is described; considerations in field testing, where model results are compared to field observations, are discussed; an assessment of the current extent of field testing for various processes and media is presented; and future field testing and data needs are enumerated.

In the past few years a variety of workshops and symposia have been held on the subjects of model verification, field validation, field testing, etc. of mathematical models for the fate and transport of chemicals in various environmental media. Following a decade of extensive model development in this area, the emphasis has clearly shifted to answering the questions "How good are these models?", "How well do they represent natural systems?", and "Can they be used for management and regulatory decision-making?" The impetus for this flurry of activity has been the recognized need for cost-effective tools, based on sound scientific principles (i.e., state-of-the-art), to assist in the performance of exposure assessments for new and existing chemical compounds and toxic wastes.

In addition to the recently published literature, much of the material for this paper was derived from technical presentations and discussions at the following workshops in which I had the opportunity to participate:

o Workshop on Verification of Water Quality Models, West Point, New York, March 1979. (1)
o Modeling the Fate of Chemicals in the Aquatic Environment Workshop, Pellston, Michigan, August 1981. (2)

0097-6156/83/0225-0151$06.25/0

    o EPA Workshop on Field Applicability Testing of Environ-
       mental Exposure Methods, Airlie House, Virginia,
       March 1982. (3)
    o EPA Exposure Assessment Workshops; Level I, Washington,
       D.C. and Level II, Atlanta, Georgia, April 1982.

Throughout these workshops the need for model testing and
validation was evident. Since models are being used, and will be
used, increasingly in the future, for exposure and risk assessments for chemicals and toxic wastes, the question of model
validity is paramount to their continued use and effectiveness
for management and regulatory decision-making. Although the
first two workshops listed above concentrated on the soil and
aquatic environments, the latter two workshops considered multimedia problems including air and groundwater concerns. Following
a brief description of each workshop, the individual conclusions
and recommendations will be synthesized and integrated into
discussions on the model validation process, model testing and
error analyses, current extent of field testing, procedures and
measures for model validation, and future needs.

Recent Model Validation Workshops

Water Quality Model Verification Workshop, West Point, NY. In
March 1979, the EPA sponsored a "National Workshop on the
Verification of Water Quality Models" (1) with the objective to
evaluate the state-of-the art of water quality modeling and make
specific recommendations for the direction and emphasis of future
modeling efforts. The participants included a cross-section of
water quality modeling experts from government, academia, industry, and private practice. The workshop was organized to address
the issues of models in decision-making, model data bases,
modeling framework and software validation, model parameter
estimation, model verification, and models as projection tools.
Each of these issues was addressed in individual work groups for
the following topical areas:
    (1) Wasteload Generation
    (2) Transport
    (3) Salinity/Total Dissolved Solids
    (4) Dissolved Oxygen/Temperature
    (5) Bacteria/Virus
    (6) Eutrophication
    (7) Hazardous Substances
Each work group produced a report of its deliberations on the
above issues for their topical area, which was published and
summarized in the workshop proceedings along with technical
discussion papers (1).

Modeling the Fate of Chemicals in the Aquatic Environment,
Pellston, Michigan. In August 1981, the fourth in the Pellston
series of technical workshops was convened to assess the current

status of modeling as an integrating mechanism for the important processes controlling chemical fate and transport. Participants included representatives from government, industry, academia, and private practice. The workshop objectives were stated as follows:

- o Analyze the state-of-the-art of environmental fate modeling considering both development and application
- o Critically evaluate existing models for predicting environmental exposure concentrations of chemicals in various aquatic systems
- o Critically examine the utility of various models as decision-making aids for their specific applications
- o Position the role of environmental fate modeling for aquatic hazard assessment, considering both regulatory applications and new product development; and
- o Develop recommendations for future research needs in environmental fate modeling (2)

The workshop was organized around four individual topics with an associated work group for synthesis and integration of the material presented: Modeling Overview - Use and Needs, Modeling (individual) Processes, State-of-the-Art and Case Study Examples, and Validation and Application Testing. The technical papers and work group reports were published in the workshop proceedings (2).

Field Applicability Testing (FAT) Workshop. In March 1982, the EPA Office of Research and Development convened a workshop with the specific objectives to (1) assess the state of knowledge on determining the field applicability of laboratory bioassay tests, toxicity studies, microcosm studies, and mathematical chemical exposure models (i.e., the extent to which these methods have been tested/compared with field data), and (2) recommend research objectives and priorities to advance the current level of field testing. Workshop attendees included representatives from EPA research laboratories, universities, and private industry. Working groups were organized with specific responsibility to assess the utility and limits of four different methods (or tools) currently used by EPA and industry for evaluating hazards posed by toxic chemicals: (1) laboratory toxicity data, (2) microcosm test data, (3) site-specific data, and (4) chemical fate and exposure model results. The Exposure Modeling Committee (3) report presented an assessment of the current extent of field model testing and recommendations for future testing efforts.

EPA Exposure Assessment Workshops - Level I and II. In April 1982, the EPA Office of Research and Development (ORD) organized two workshops designed to assess and identify current techniques (i.e., data, protocols, predictive models) used in performing exposure assessments, enumerate technical information gaps, and recommend prioritized research topics to satisfy current and anticipated needs. The Level I workshop was comprised of

technology users (for performing exposure assessments) in the EPA
program offices, ORD modelers and outside experts from academia,
industry, and consulting firms. The goal was to join together
the users and developers of exposure assessment techniques to
identify current methods, technology gaps, and future needs. The
subsequent Level II workshop, composed of model developers and
scientists, deliberated on the specific technical issues identified in the Level I workshop and developed a ranking of specific
research activities needed to satisfy current and projected
exposure assessment needs. The material presented in this paper
for this workshop is based on the observations of the author and
preliminary workshop proceedings.

The Model Testing/Validation Process

Part of the confusion surrounding the model testing and validation process is largely because different meanings have been
attached to the terms calibration, verification, validation,
and post-audit in the technical literature. As a result of the
Pellston conference, I have adopted the following relationship
among these terms:

$$\left.\begin{array}{l} \text{Calibration} \\ \text{Verification} \\ \text{Post-Audit} \end{array}\right\} = \text{Model Testing or Validation}$$

Thus, the process of model testing and validation (considered
synonymous) should ideally include the steps of calibration
(if necessary), verification, and post-audit analyses. I
indicate "ideally" because in many applications existing data
will not support performance of all steps. In chemical fate
modeling, chemical data for verification is often lacking and
post-audit analyses are rare (unfortunately) for any type of
modeling exercise.
     Figure 1 presents an overview of the model testing/valida-
tion process as developed at the Pellston workshop. A distinction
is drawn between validation of empirical versus theoretical models
as discussed by Lassiter (4). In reality, many models are
combinations of empiricism and theory, with empirical formulations
providing process descriptions or interactions lacking a sound,
well-developed theoretical basis. The importance of field data
is shown in Figure 1 for each step in the model validation
process; considerations in comparing field data with model
predictions will be discussed in a later section.
     The model construction check is needed to confirm the
correct structure and operation of the model algorithms over the
range of conditions and model parameters expected.
     Calibration is probably the most misunderstood of all the
model validation components. Calibration is the process of
adjusting selected model parameters within an expected range
until the differences between model predictions and field
observations are within selected criteria for performance (to be

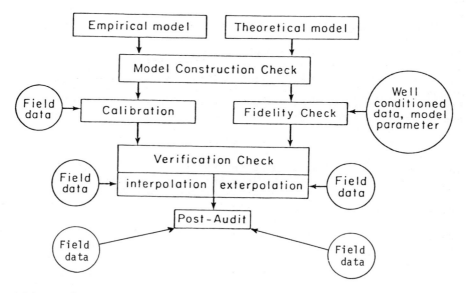

Figure 1.   The process of model testing/validation. (Reproduced
with permission from Ref. 2.)

discussed later). For certain disciplines, such as air transport
models and hydromechanic models for tsunami propagation or
hurricane surge, the underlying theory is sufficiently well-
developed that calibration is minimal or not needed (5).

However, for all operational, empirical models (or portions
thereof), including most chemical fate models, calibration is
usually needed and highly recommended. The advocates of calibration (including this author) argue that calibration is needed
to "account for spatial variations not represented by the model
formulation; functional dependencies of parameters that are
either non-quantifiable, unknown, and/or not included in the
model algorithms; or extrapolation of laboratory measurements of
parameters to natural field conditions" (6). On the other hand,
the opponents argue that calibration is essentially a curve-
fitting procedure using the numerous degrees-of-freedom of the
model to match observed and predicted values. It is clear that
the need for calibration increases the user effort and data
required to appropriately apply a model. However, any model can
be operated without calibration depending on the extent to which
critical model parameters (usually refined through calibration)
can be estimated from past experience and other data. In the
area of pesticide runoff modeling, Lorber and Mulkey have shown
that so-called "calibration-independent" (empirical, in this
case) models produced their "'best' results...only after some
deliberation and reassignment of initial parameter estimates" (7).
In many cases, calibration becomes an expedient alternative
when the calibration-independent model fails to satisfy acceptance
criteria with the originally estimated parameter values.

Verification is the complement of calibration; model
predictions are compared to field observations that were not used
in calibration or fidelity testing. This is usually the second
half of split-sample testing procedures, where the universe of
data is divided (either in space or time), with a portion of the
data used for calibration/fidelity check and the remainder used
for verification. In essence, verification is an independent
test of how well the model (with its calibrated parameters) is
representing the important processes occurring in the natural
system. Although field and environmental conditions are often
different during the verification step, parameters determined
during calibration are not adjusted for verification.

Post-Audit Analyses are the ultimate tests of a model's
predictive capabilities. Model predictions for a proposed alternative are compared to field observations following implementation
of the alternatives. The degree to which agreement is obtained
based upon the acceptance criteria reflects on both the model
capabilities and the assumptions made by the user to represent
the proposed altenative. Unfortunately, post-audit analyses
have been performed in few situations.

Model Testing and Error Analysis

The process of field validation and testing of models was
presented at the Pellston conference as a systematic analysis
of errors (6). In any model calibration, verification or validation effort, the model user is continually faced with the
need to analyze and explain differences (i.e., errors, in this
discussion) between observed data and model predictions. This
requires assessments of the accuracy and validity of observed
model input data, parameter values, system representation, and
observed output data. Figure 2 schematically compares the model
and the natural system with regard to inputs, outputs, and sources
of error. Clearly there are possible errors associated with each
of the categories noted above, i.e., input, parameters, system
representation, output. Differences in each of these categories
can have dramatic impacts on the conclusions of the model validation process.

Input Errors. Errors in model input often constitute one of the
most significant causes of discrepancies between observed data
and model predictions. As shown in Figure 2, the natural system
receives the "true" input (usually as a "driving function")
whereas the model receives the "observed" input as detected by
some measurement method or device. Whenever a measurement is made
a possible source of error is introduced. System inputs usually
vary continuously both in space and time, whereas measurements are
usually point values, or averages of multiple point values, and
for a particular time or accumulated over a time period. Although
continuous measurement devices are in common use, errors are still
possible, and essentially all models require transformation of
a continuous record into discrete time and space scales acceptable
to the model formulation and structure.
    A classic example of input errors in watershed hydrologic
modeling is the use of point rainfall measurements to estimate
the effective rainfall over a watershed area. Since rainfall is
the driving time series in hydrologic modeling, any errors in
the input rainfall, i.e., misrepresentation of effective watershed
rainfall, will likely propagate through the simulation and thus
have a critical effect on model calibration and verification.
Donigian, et. al., (8) describe an example of this type of error
in an application of the Hydrologic Simulation Program FORTRAN
(9) for analysis of agricultural best management practices in a
5000-hectare Iowa watershed. Due to limited precipitation data,
it was necessary to use data from different gage locations for
calibration and verification. Comparison of model runs for the
same time period using the two rainfall records (i.e., one within
the watershed and the other 10 kilometers east) resulted in a
35% change in mean flow, although the rainfall volumes differed
by less than 5%. Verification procedures were subsequently
modified to provide a more consistent basis for evaluating model

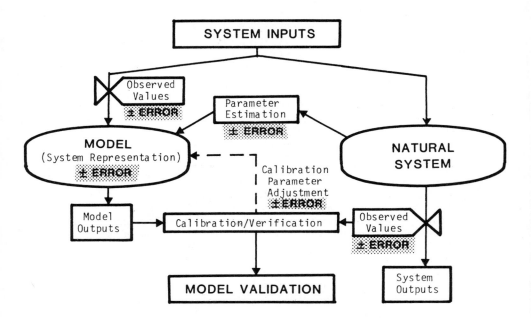

Figure 2. Model vs. natural system: inputs, outputs, and errors.
(Reproduced with permission from Ref. 2.)

performance. Similar discrepancies were noted by Troutman in
analyzing the effects of spacial variability of precipitation on
runoff predictions using a USGS model applied to a Texas watershed (10).

Another example of possible input errors especially critical
for chemical fate modeling is errors associated with input
loadings of chemicals. For a watershed modeling effort, these
errors would be associated with chemical input via rainfall or
dry deposition; chemical land application rates, methods, and
timing; and point discharges of chemicals. For pesticides and
other more exotic chemicals, rainfall and dry deposition can
usually be ignored unless drift from aerial applications in
neighboring air sheds is significant.

Assumptions on chemical application rates, methods, and
timing were shown to have major impact on resulting concentrations and loads measured at the watershed outlet (8). Application
rates determined by an inventory of farmers can include substantial error since many farmers may not follow label recommendations. Surface versus subsurface application methods affect the
susceptibility to removal by surface runoff, erosion, and/or
subsurface flow components. Timing of applications is especially
critical for fast decaying pesticides since the majority of the
pesticide loss by runoff occurs in the first few storm events
following application. Consequently, the amount and timing of
application, e.g., before or after a specific event, has a
major impact on how much loss occurs during the event.

For an aquatic model of chemical fate and transport, the
input loadings associated with both point and nonpoint sources
must be considered. Point loads from industrial or municipal
discharges can show significant daily, weekly, or seasonal
fluctuations. Nonpoint loads determined either from data or
nonpoint loading models are so highly variable that significant
errors are likely. In all these cases, errors in input to a
model (in conjunction with output errors, discussed below) must
be considered in order to provide a valid assessment of model
capabilities through the validation process.

System Representation Errors. System representation errors refer
to differences in the processes and the time and space scales
represented in the model, versus those that determine the
response of the natural system. In essence, these errors are the
major ones of concern when one asks "How good is the model?".
Whenever comparing model output with observed data in an attempt
to evaluate model capabilities, the analyst must have an understanding of the major natural processes, and human impacts, that
influence the observed data. Differences between model output and
observed data can then be analyzed in light of the limitations
of the model algorithm used to represent a particularly critical
process, and to insure that all such critical processes are
modeled to some appropriate level of detail. For example, a

lake model that simulates dissolved oxygen (DO) without
simulating biological components cannot be expected to reproduce
observed DO values during algae bloom conditions. Similarly,
for chemical transport modeling a compound that even partially
adsorbs to sediment particles cannot be accurately represented
without modeling sediment transport and sediment-chemical
interactions. A model without algorithms for these processes
may be quite appropriate for completely soluble (non-adsorbing)
compounds, but it would not be applicable to many compounds
that undergo adsorption and desorption.
    Although the existence or absence of a particular process
can often be determined from observed data, an assessment of how
well an algorithm represents the process is often difficult to
make due to observation errors, natural variations in field
data, and lack of sufficient data on individual component
processes. In such circumstances, model validity must be
inferred or possibly based on comparisons with laboratory data
obtained under controlled conditions. Often laboratory data
provide the basis for developing an algorithm since field data
are so much more difficult and expensive to collect and interpret. Examples of system representation errors and their
analysis were presented at the Pellston workshop (6).

Parameter Errors. Parameter errors are derived primarily from
the inability to accurately measure and predict many of the
parameters that characterize the natural system and, for chemical
fate modeling, the relevant chemical processes under field
conditions. Errors are associated both with parameter values
obtained from actual measurements and from calibration. Due to
natural variations in topography, soil characteristics, crop
cover densities, etc., a single parameter value or time-variable
parameter function (e.g., crop cover) will always have some
associated error. The goal is to obtain measured parameter
values that, to the extent possible, represent mean or average
conditions for the natural system. If the specific parameter
values demonstrate significant variations, then further segmen-
tation of the system representation may be needed to develop
regions with relatively uniform characteristics. Such is the
case for different crop types and vastly different soils
characteristics for watershed modeling. This is also related
to the spatial scales required to adequately represent the
system.
    Parameters for which measured values are not clearly defined
or readily available are often determined through calibration with
observed data. In watershed chemical fate modeling, calibration
has traditionally been associated with hydrologic parameters (e.g.,
erodibility coefficients, scour and deposition rates) because the
required flow and sediment data are often available. Although
initial parameter values can always be estimated, calibration is
usually recommended to account for local and spatial variations.

Chemical parameters (e.g., partition coefficients, decay rates, temperature and moisture effects) are not usually considered as calibration parameters because they can be measured in a laboratory; moreover, calibration is usually not possible due to lack of observed data. However, most scientists will agree that extrapolation of laboratory parameter measurements to field conditions is a risky assumption. If observed chemical data are available, refinement of initial chemical parameters through calibration should be considered. Errors in calibration-derived parameter values are often a function of how much calibration was performed or errors in system inputs and/or outputs. In many modeling efforts, conscientious model users will often overrun the calibration budget because of the natural tendency to continue to make calibration runs in an effort to minimize discrepancies between simulated and observed values. Parameter errors associated with calibration are more often a result of missing and/or erroneous data either as system inputs or outputs.

Errors in system output measurements can produce calibration errors because the model user will be attempting to calibrate against inaccurate or missing data. Errors associated with system outputs are discussed below.

Output Errors. Output errors are analogous to input errors; they can lead to biased parameter values or erroneous conclusions on the ability of the model to represent the natural system. As noted earlier, whenever a measurement is made, the possibility of an error is introduced. For example, published U.S.G.S. stream-flow data often used in hydrologic models can be 5 to 15% or more in error; this, in effect, provides a tolerance range within which simulated values can be judged to be representative of the observed data. It can also provide a guide for terminating calibration efforts.

Output errors can be especially insidious since the natural tendency of most model users is to accept the observed data values as the "truth" upon which the adequacy and ability of the model will be judged. Model users should develop a healthy, informed scepticism of the observed data, especially when major, unexplained differences between observed and simulated values exist. The FAT workshop described earlier concluded that it is clearly inappropriate to allocate all differences between predicted and observed values as model errors; measurement errors in field data collection programs can be substantial and must be considered.

A dramatic example of this type of error was discussed by Donigian, (6) at the Pellston workshop based on the Iowa study described earlier (8); Figure 3 shows the calibration (top figure, 1978 data) and verification (bottom figure, 1978) results. A simulated alachlor concentration value of greater than 0.1 mg/l occurred on May 27, 1978, (top figure) whereas the observed

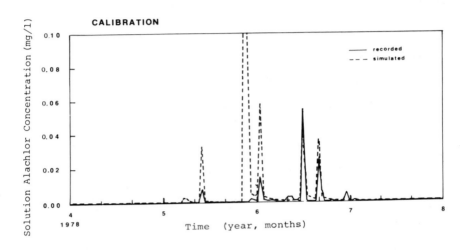

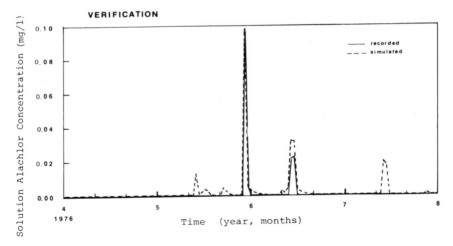

Figure 3.   Calibration and verification results for solution
Alachlor concentration at Traer, Iowa. (Reproduced with
permission from Ref. 2.)

value is zero. Since May is the month of alachlor applications
in Central Iowa it is logical to expect significant alachlor
concentrations for even minor runoff events during this period.
Coincidentally, a storm of comparable rainfall and flow occurred
during the verification period on May 29, 1976, (bottom graph in
Figure 3) producing a maximum alachlor concentration of about
0.1 mg/l, which was equal to the simulated value. Further
investigation showed that problems in communications led to
missing samples on both May 27 and June 1, 1978.
   In comparing the May storms of 1978 and 1976, clearly the
simulated concentration values in Figure 3 are more representative
of what actually occurred than the observed values. This is not
meant to be a criticism of the sampling program but an indication
of how errors in observed data can exist and impact the model
validation process.
   In addition to outright errors in observed data, the model
user needs to assess how representative the data are considering
sampling methods and locations, and in view of subsequent manipulation of the data, for example, to estimate chemical loads
based on the flow and concentration values. While observed data
are measurements at a particular point on a stream, the model
simulation represents an average concentration in an assumed,
completely-mixed section of the stream. Also, the method of
sample collection, e.g., depth integrated, cross-sectionally
integrated, or grab samples - should be considered in terms of
obtaining a representative sample.
   Comparisons between observed data and model predictions
must be made on a consistent basis, i.e., apples with apples
and oranges with oranges. Since models provide a continuous
timeseries, any type of statistic can be produced such as daily
maximums, minimums, averages, medians, etc. However, observed
data are usually collected on infrequent intervals so only
certain statistics can be reliably estimated. Validation of
aquatic chemical fate and transport models is often performed
by comparing both simulated and observed concentration values
and total chemical loadings obtained from multiplying the flow
and the concentration values. Whereas the model supplies flow
and concentration values in each time step, the calculated
observed loads are usually based on values interpolated between
actual flow and sample measurements. The frequency of sample
collection will affect the validity of the resulting calculated
load. Thus, the model user needs to be aware of how observed
chemical loads are calculated in order to assess the veracity of
the values.
   Although this section has emphasized possible errors in the
observed data, the above considerations should not be used as a
crutch to support an invalid or inaccurate model. In truth, in
most circumstances the observed data are our best real indication
of system behavior. Combining model simulations with an informed
skepticism of both model and the observed data can lead to a
better overall understanding of modeling natural systems.

Current Status of Field Testing

The Exposure Modeling Work Group of the FAT Workshop identified
transport and transformation processes incorporated in existing
exposure models, evaluated the current extent of field applicability testing, and enumerated specific research and field data
collection needs. The specific media and environmental systems
considered included air, runoff, streams, lakes, estuaries, and
soil systems (unsaturated and saturated). The Work Group
concluded that the current extent and/or adequacy of model field
testing could only be assessed with respect to the model accuracy
required for specific types of regulatory problems. Screening
and site-specific assessments for all media were identified as
the most likely problems expected under current and future
regulatory conditions. Although specific precision and accuracy
requirements could not be defined, these two levels of assessment
were characterized as follows:
  o screening - screening level modeling (e.g., chemical
     pre-manufacturing notice evaluations) is usually
     accomplished by far-field models and accuracy is
     expected to be within one order of magnitude.
  o site-specific - more detailed modeling is required for
     site-specific problems (e.g., waste load allocations,
     hazardous waste system siting) and accuracy is expected
     to be within a factor of 2-4 for many situations and
     within a factor of less than 2 for some situations. (3)
The current extent of field testing was evaluated for each level
since the testing and data requirements for site-specific assessments would be significantly more stringent than for screening
purposes.
     Tables I and II present the results of the Work Group
discussions for the screening and site-specific level models,
respectively. The assessment in these tables is based on a
ranking scale between 0 and 100; 0 indicates situations where
no testing has been attempted and 100 identifies areas where
extensive testing has been completed with sufficient post-audits
to validate the predictive capability of relevant models. The
scores can also be interpreted to mean the extent to which
additional field testing would improve our understanding of how
well the models represent natural systems. It is important to
note that the scores do not indicate model accuracy per se; they
show the degree to which current field testing has been able to
identify or estimate model accuracy.
     The tables were designed to encompass processes included in
most models of the various media of concern. Although selected
processes are not rigorously defined for each media (e.g.,
sorption/desorption in air refers to toxicant-particulate interactions), the goal was to provide a concise ranking table for
each level of analysis.

Table I. CURRENT EXTENT OF FIELD TESTING FOR SCREENING LEVEL MODELS (3)
100 = Model thoroughly field tested
0 = Model not tested

MEDIA

MODELED PROCESS	Air	Watersheds Runoff	Watersheds Streams	Lakes	Estuaries	Soil Systems Unsat. Zone	Soil Systems Sat. Zone
Dispersion-Diffusion	90	--	100	100	100	50	75
Advection	75	100	100	100	100	100	100
Intermedia Transfers							
-sorption/desorp.	75	80	80	80	80	80	80
-volatilization	10	50	100	80	80	70	50
-wet/dry deposition							
-to land	50	--	--	--	--	--	--
-to water	90	--	--	--	--	--	--
Sediment/Particulate Transport	75	60	60	90	60	--	--
Transformation Processes							
-chemical	75	--	80	80	80	10	10
-biological	--	--	10	10	10	10	10
-lumped parameter	--	--	--	--	--	60	25

FATE OF CHEMICALS IN THE ENVIRONMENT

Table II. CURRENT EXTENT OF FIELD TESTING FOR SITE-SPECIFIC MODELS (3)

100 = Model thoroughly field tested
0 = Model not field tested

MEDIA

MODELED PROCESS	Air	Watersheds		Lakes	Estuaries	Soil Systems	
		Runoff	Streams			Unsat. Zone	Sat. Zone
Dispersion-Diffusion	80	--	90	80	80	10	25
Advection	25	75	100	80	80	15	30
Intermedia Transfers							
-sorption/desorp.	50	40	40	40	20	20	30
-volatilization	10	15	80	20	20	15	15
-wet/dry deposition							
-to land	10	--	--	--	--	--	--
-to water	60	--	--	--	--	--	--
Sediment/Particulate							
Transport	60	30	20	30	10	--	--
Transformation Processes							
-chemical	45	--	80	80	80	10	10
-biological	--	--	10	10	10	10	10
-lumped parameter	--	--	--	--	--	60	25

As expected, the tables clearly show that the transport processes have been more thoroughly tested for all media than related transformation processes. Also, chemical transformations have been more extensively tested than biological transformations. Both transport and transformation processes for soil systems ranked consistently lower than the same processes in water systems. The site-specific assessment reflected in Table II is clearly supported by the results of the earlier West Point and Pellston workshops and the subsequent EPA Exposure Assessment workshops. Sediment/particulate transport (especially for non-cohesive sediments) and sediment-contaminant-water interactions were listed as major model development and testing needs, with particular emphasis on testing existing models as a means to further development and refinement. Testing of sediment transport and erosion models should consider the importance of these processes for modeling toxicant transport. Mixing processes, both at the land surface (primarily for the availability of chemicals to runoff) and at the stream bed, need further investigations in terms of physical, chemical, and biological (i.e., bioturbation) processes; the common assumption of an active zone or depth of mixing should be further refined.

In all of the workshops, but especially in the FAT and Exposure Assessment workshops, the need for better understanding and model representation of soil systems, including both unsaturated and saturated zones, was evident. This included the entire range of processes shown in Table II, i.e., transport, chemical and biological transformations, and intermedia transfers by sorption/desorption and volatilization. In fact, the Exposure Assessment workshop (Level II) listed biological degradation processes as a major research priority for both soil and water systems, since current understanding in both systems must be improved for site-specific assessments.

Procedures and Measures for Model Validation

Despite a recent emphasis on model testing and validation throughout the literature, unified and accepted procedures and measures for model validation do not exist at the present time. At the West Point workshop (1) a clear consenus was evident throughout all the work groups that quantitative measures of model performance (both calibration and verification) should be an integral part of the modeling efforts and used to supplement qualitative assessments. These concepts were re-confirmed at the Pellston workshop where it was noted that development of a statistical foundation for model testing was needed to provide accepted methodologies for the modeling community. The goal is to provide definitive answers to the common questions asked by the manager or decision-maker, i.e., "How good is the model?" and "What is the level of confidence that we can place on your results?" (11).

Although procedures for model testing are often problem and model specific, the FAT workshop (3) identified three general categories in common use for field testing:
1. Model parameter estimation by laboratory, microcosm, or pilot plant studies followed by field application. Many theoretical models will require laboratory measures of parameters, such as rate constants or partition coefficients, not easily measured in the field.
2. "Split-sample" field testing involving calibration and verification on separate data sets, often for different time periods at one site. This approach has been used widely in hydrologic modeling and has general utility for model testing.
3. Site to site extrapolation of model results involving model calibration at one site and subsequent testing against data collected at another site. This is analogous to the classic case of the "ungaged" watershed problem in hydrology, where estimates at an ungaged site are derived from model applications at locations further up or downstream, or from a neighboring site.

These three procedures are often combined in various ways depending on data availability, model structure, and modeling purposes. For example, transport processes may often be calibrated and verified on available data, while the transformation process parameters may be derived from laboratory measurements and applied without calibration.

The greatest need in model performance testing and validation is clearly the use of quantitative measures to describe comparisons of observed and predicted values. As noted above, although a rigorous statistical theory for model performance assessments has yet to be developed, a variety of statistical measures has been used in various combinations and the frequency of use has been increasing in recent years. The FAT workshop (3) identified three general types of comparisons that are often made in model performance testing:
1. Paired-data performance, involving comparison of predicted and observed values for exact locations in time and space. This may be a more rigorous test than needed for many purposes; timing differences can have severe impacts on the statistical comparison.
2. Time and space integrated, paired-data performance. Spacially and/or temporally integrated data can be compared to analogous model predictions, such as daily or monthly averages or totals. This can circumvent some of the timing problems noted in (1).
3. Frequency domain performance, involving comparison of cumulative frequency distributions of the observed data and model predictions. In many situations, considering the various sources of error discussed earlier, it may

be less important to match specific events or time
periods than to characterize the general response of
the system by its frequency distribution. This is a
common practice in hydrology where flow frequency
duration curves are used.
Statistical measures for the paired-data, and integrated paired-
data performance tests noted above are essentially identical.
They include simple statistics (e.g., sums, means, standard
deviations, coefficient of variation), error analysis terms
(e.g., average error, relative error, standard error of estimate),
linear regression analysis, and correlation coefficients.
Thomann (11) discusses various model verification measures and
demonstrates their application to eutrophication model results.
Studies by Ambrose and Roesch (12), Hartigan et. al. (13),
Young and Alward (5), and Lorber and Mulkey (7) demonstrate the
use of these statistical measures as quantitative assessments of
model performance for a wide range of models.
     Frequency domain performance has been analyzed with goodness-
of-fit tests such as the Chi-square, Kolmogorov-Smirnov, and
Wilcoxon Rank Sum tests. The studies by Young and Alward (14)
and Hartigan et. al. (13) demonstrate the use of these tests for
pesticide runoff and large-scale river basin modeling efforts,
respectively, in conjunction with the paired-data tests. James
and Burges (16) discuss the use of the above statistics and some
additional tests in both the calibration and verification phases
of model validation. They also discuss methods of data analysis
for detection of errors; this last topic needs additional
research in order to consider uncertainties in the data which
provide both the model input and the output to which model
predictions are compared.
     The use of statistical tests to characterize model perfor-
mance has increased markedly within the last few years as
demonstrated by the references cited above; many other examples
are in the literature. The current time is appropriate for
integration of existing techniques into a unified framework of
guidelines, procedures and quantitative measures for model
testing and validation.

Future Needs

Future needs in support of model validation and performance
testing must continue to be in the area of coordinated, well-
designed field data collection programs supplemented with directed
research on specific topics. The FAT workshop produced a listing
of the field data collection and research needs for the air,
streams/lakes/estuaries, and runoff/unsaturated/saturated soil
media categories, as follows:
     o air: long-range transport phenomena, dry deposition over
               complex (inhomogeneous) terrain, dry deposition as
               affected by chemical species, wet deposition,

               aerosol formation, and toxic chemical transforma-
               tion.
o streams/lakes/estuaries: contaminant transformations in
        sediments, transfers between bottom sediments and
        overlying water, sediment and sorbed pollutant
        transport and transformation rates and kinetics.
o runoff/unsaturated/saturated soil media:
      - runoff/unsaturated/stream watershed systems -
        sediment transport, adsorption/desorption, and
        transformations
      - unsaturated/saturated flow systems - transforma-
        tions, volatilization, adsorption/desorption,
        dispersion/diffusion, and leaching
      - saturated flow systems - transformations, adsorp-
        tion/desorption, dispersion/diffusion, and
        advection.
Specific research investigations into sediment/particulate
transport, sediment/water/contaminant interactions, soil (unsaturated and saturated) contaminant fate and transport, and biological degradation processes were identified as priorities by the
Exposure Assessment workshops.

Finally, the need exists for unified and consistent procedures and/or guidelines for model performance testing and
validation, supplemented with rigorous and accepted statistical
measures that allow consideration of data error/uncertainties.
Perhaps, current efforts by the American Society of Civil
Engineers and The American Society of Testing and Materials to
establish modeling and model testing guidelines will begin to
satisfy this need.

Literature Cited

1. Thomann, R.V.; Barnwell, T.O. Jr.,"Workshop on Verification
    of Water Quality Models",U.S. EPA, Athens, GA., EPA-600/9-80-
    016, 1980.
2. Dickson, K.L.; Maki, A.W.; Cairns, J. Jr., "Modeling the Fate
    of Chemicals in the Aquatic Environment"; Ann Arbor Science
    Publishers; Ann Arbor, MI., 1982; p. 413.
3. Exposure Modeling Committee Report. Testing for the Field
    Applicability of Chemical Exposure Models. Proc. Workshop
    on Field Applicability Testing. U.S. EPA, Athens, GA.,
    1982.
4. Lassiter, R.R., in "Modeling the Fate of Chemicals in the
    Aquatic Environment"; Ann Arbor Science Publishers; Ann
    Arbor, MI., 1982; p 287-301.
5. Young, G.K.; Alward, C.L., Calibration and Testing of Pesti-
    cide Transport Models. Presented at Iowa State University
    Conference on Ag. Management and Water Quality. 1981.

6. Donigian, A.S. Jr., in "Modeling the Fate of Chemicals in the Aquatic Environment"; Ann Arbor Science Publishers; Ann Arbor, MI., 1982; p. 303-323.

7. Lorber, M.N.; Mulkey, L.A., J. Env. Qual. 1982, 11, (3), 519-528.

8. Donigian, A.S. Jr.; Imhoff, J.C.; Bicknell, B.R., "Modeling Water Quality and the Effects of Agricultural Best Management Practices in Four Mile Creek, Iowa." U.S. EPA Athens, GA., 1983.

9. Johanson, R.L.; Imhoff, J.C.; Davis, H.H. Jr.; Kittle, J.L. Jr.; Donigian, A.S. Jr., User's Manual for Hydrological Simulation Program - FORTRAN (HSPF). Release No. 7 U.S. EPA, Athens, GA., 1981.

10. Troutman, B.M., in "Statistical Analysis of Rainfall-Runoff" Singh, V.P. Ed.; W.R. Pub. 1982; p 305-314.

11. Thomann, R.V., in "Workshop on Verification of Water Quality Models"; Thomann, R.V.; Barnwell, T.O., Ed.; U.S. EPA, Athens, GA., EPA600/9-80-016, 1980, p 37-61.

12. Ambrose, R.B. Jr.; Roesch, S.E., ASCE EE1, 1982; p 51-71.

13. Hartigan, J.P.; Quasebarth, T.F.; Southerland, E., Proc. Stormwater and W.Q. Model Users Group Meeting, 1982.

14. James, L.D.; Burges, S.J., "Hydrologic Modeling of Small Watersheds"; Haan, C.T.; Johnson, H.P.; Brakensiek, D.L., Ed. ASAE. Mono. #5, 1982; Chapter 11, p 437-472.

RECEIVED April 15, 1983.

MODEL PARAMETERS

Application of Fugacity Models to the Estimation of Chemical Distribution and Persistence in the Environment

DONALD MACKAY, SALLY PATERSON, and MICHAEL JOY

University of Toronto, Department of Chemical Engineering and
Applied Chemistry, Toronto, Ontario, Canada M5S 1A4

The roles of mathematical models for predicting the
likely behavior of chemicals in real and evaluative
environments are discussed, and it is suggested
that more consideration should be given to defining
acceptable levels of model complexity. The concepts underlying a series of fugacity models are
described and illustrated by applying the models to
(i) an assessment of the behavior of a trichlorobiphenyl with four fugacity models of an evaluative
lake environment as an illustration of various
levels of complexity (ii) an assessment of the
relative behavior of mono, di, tri and tetra
chlorobiphenyls in the same environment as an
illustration of the effect of changing chemical
properties on behavior and (iii) a description of
trichlorobiphenyl behavior in a lake similar to
Lake Michigan using the QWASI (Quantitative Water
Air Sediment Interaction) fugacity model. It is
concluded that evaluative models can generate
behavior profile information of value for hazard
assessment purposes by integrating data on partitioning, reaction, advection, and inter-phase
transport. By applying the same concepts and
equations to models of real environments and vali-
dating them, the evaluative and real modeling
efforts become mutually supportive and the credibility of both is increased.

In a series of recent papers (1 - 4), we have advocated the
use of the fugacity concept as an aid to compartmental modeling
of chemicals which may be deliberately or inadvertantly discharged
into the environment. The use of fugacity instead of concentration may facilitate the formulation and interpretation of environmental models; it can simplify the mathematics and permit processes which are quite different in character to be compared

quantitatively in order that the dominant processes can be identified. Fugacity is easily conceived as an escaping tendency or
pressure with units of pressure (eg. Pa). In this paper, we
review briefly the underlying concepts of fugacity modeling,
discuss its application to real and evaluative environments, and
demonstrate that the models may be applied at various selected
levels of complexity.

We believe that one key to successful environmental compartment modeling is to identify first the required or acceptable
level of model complexity, then include the dominant processes in
the model followed by others in order of decreasing importance
until the desired level of complexity is achieved. Other, less
important processes are ignored. The modeler's defence to criticism that a process has been ignored is then clear - inclusion
would exceed acceptable complexity. The difficult and possibly
contentious step is to rank processes in order of importance. We
suggest that this ranking can rarely be done a priori, it is
usually the result of trial and error. Indeed the art of environmental modeling lies in the ability of the modeler to concep-
tualize the problem, identify a sufficient number of dominant
processes and then write reasonable descriptive equations for each
process. Manipulating the equations to obtain a solution is
usually the least difficult task.

Real environments are usually characterized by inherent complexity and a corresponding inadequacy of detailed understanding
of properties and process rates. Chemical concentrations vary
temporally and spatially. The bulk movement of air, water, suspended soils and biota is irregular and difficult to describe
numerically. Reactions are numerous, interactive and may vary
diurnally and seasonally. Faced with this complexity, the model-
ler's response is to average properties and processes, assume
homogeneity instead of heterogeneity, and generally simplify the
system until it becomes tractable. Accomplishing this task for a
real environment requires considerable effort and insight with the
result that the modeler may have little intellectual energy (or
funds) left for interpreting the behavior of the chemical, as
distinct from the behavior of the environment.

A very significant advance was made by Baughman and Lassiter
(5) when they suggested using evaluative environments for eluci-
dation of the environmental behavior of chemicals. This led to
the EXAMS model (6), the studies of selected chemicals by Smith
et al (7, 8), the development of "Unit Worlds" by Neely and Mackay
(9) and Mackay and Paterson (2), and the incorporation of similar
Unit Worlds into hazard assessment by Schmidt-Bleek et al (10).
The evaluative approach frees the modeler from concerns about
environmental identification and enables all attention to be
focused on the chemical's behavior. An unfortunate consequence
is that direct validation is not possible; thus there may be
reluctance to accept the conclusions. Perhaps this reluctance may
best be alleviated by demonstrating that the evaluative model

process can be applied successfully to microcosms, to well controlled outdoor environments such as small ponds or agricultural
plots, or to rivers or lakes.

Modeling of evaluative and real environments should be viewed
as complementary. Evaluative models are particularly suitable for
assessment of new chemicals, for comparing chemicals, and for
obtaining general chemical behavior profiles. Real models are
obviously best used for elucidating the actual or potential nature
of contamination situations and remedial actions. The use of
similar or identical calculation techniques in both is very
desirable since success in the real case may lead to greater
credibility in the evaluative case.

Fugacity Models

An attractive feature of the fugacity models is that they
can be applied at various levels of complexity, depending on the
perceived modelling need and the availability of data. The
determinants of complexity are believed to be as follows.
1. Number of compartments considered.
2. If phase equilibrium is assumed between some or all
   compartments.
3. If degradation reactions are included.
4. If advection processes are included.
5. If steady state is assumed or time dependence of
   concentration and emissions is included.

A fugacity level I calculation may be 6 compartment equilibrium, no reaction, no advection, steady state; a level II may
be equilibrium, with reaction and advection, steady state; level
III may be non equilibrium, with reaction and advection, steady
state, and level IV and EXAMS are non equilibium, with reaction and
advection, unsteady state.

With models being formulated by many independent groups, it
is inevitable that comparisons will be made in the hope of identifying the better or more useful models. Comparison between models
of different classes is usually not meaningful.

Equilibrium. Equilibrium between compartments can be expressed either as partition coefficients K_{ij} (i.e. concentration ratio
at equilibrium) or in the fugacity models as fugacity capacities
Z_i and Z_j such that K_{ij} is Z_i/Z_j, the relationships being depicted
in Figure 1. Z is defined as the ratio of concentration C
(mol/m^3) to fugacity f (Pa), definitions being given in Table I.

An advantage of the fugacity capacity approach is that for N
compartments N values of Z are defined while there may be $N(N-1)/2$
partition coefficients. Using Z values the partitioning properties between two phases are attributed independently to each
phase. It is possible to assign (accidentally) three inconsistent
partition coefficients between air, soil and water but the three Z
values are inherently consistent.

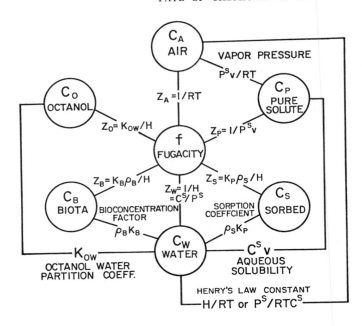

Figure 1.   Relationships between fugacity capacities and
partition coefficients.   See Table 1 for symbol definitions.

Table I.   Definition of Fugacity Capacities

Compartment	Definition of Z (mol/m³ Pa)	
Air	$1/RT$ R=8.314 Pa m³/mol K T=Temp. (K)	
Water	$1/H$ or C^S/P^S C^S = aqueous solubility (mol/m³)	
		P^S = vapor pressure (Pa)
		H = Henry's law constant (Pa m³/mol)
Solid Sorbent (e.g. soil sediment, particles)	$K_p \rho_s /H$	K_p = partition coeff. (L/kg)
		ρ_s = density (kg/L)
Biota	$K_B \rho_B /H$	K_B = bioconcentration factor (L/kg)
		ρ_B = density (kg/L)
Pure Solute	$1/P^S v$	v = Solute molar volume (m³/mol)

Reactions. Reactions are expressed by first order equations in chemical concentration (rate constant $k_i h^{-1}$) such that the rates of processes such as hydrolysis, oxidation, photolysis, or biolysis can be combined by adding the k terms to yield a total rate constant k_T.

$$\text{rate} = k_1 C + k_2 C + k_3 C \text{ etc} = C \sum k_i = C k_T \text{ mol/m}^3 \text{h}$$

This rate can also be expressed in terms of fugacity f for a compartment of volume Vm³ as

$$\text{Rate (mol/h)} = V C k_T = V Z k_T f = D_R f$$

The D_R term or group ($V Z k_T$) has units of mol/hPa and can be viewed as a rate of loss of chemical from the compartment (by reaction) per unit of fugacity.

Advection. Advection to and from a compartment of volume V at flow rate G m³/h by, for example, air or water flow can be expressed as a pseudo first order rate process with a rate constant k_A equal to G/V. The rate is also given by

$$\text{Rate (mol/h)} = GC = V C k_A = V Z k_A f = GZf = D_A f$$

The D_A term or group (GZ) again has units of mol/hPa and is a rate of loss or gain of chemical from the compartment (by advection) per unit of fugacity.

Interphase Diffusion. When interphase transport rates are characterized it can be shown that the diffusion rate between two compartments i and j can be expressed as (3)

$$\text{Rate (mol/h)} = D_{ij}(f_i - f_j)$$

Where D_{ij} can be calculated from mass transfer coefficients or an uptake half-time. For example, for air-water exchange D_{AW} is given by

$$D_{AW} = A/(1/K_A Z_A + 1/K_W Z_W)$$

where A is the surface area (m²) and K_A and K_W are the air and water phase mass transfer coefficients (m/h). For uptake by fish (subscript F) a half-time τ (h) can be used.

$$D_{FW} = 0.69 V_F Z_F / \tau$$

Inter-phase diffusion processes can be viewed as the net result of two counter-processes with rates of $D_{ij} f_i$ and $D_{ij} f_j$.

Interphase Material Transfer. In some cases there is uni-directional bulk transfer of material and associated chemical between compartments (e.g. sediment deposition or atmospheric particle fallout) in which case the rate is given by an expression similar to that for advection in which G_B (m³/h) is the rate of transfer of the material namely

$$\text{Rate (mol/h)} = G_B C = G_B Zf = D_B f$$

again D is a loss or gain coefficient and has units of mol/hPa.

Time Dependence. The implications of introducing time dependence are obvious. Steady state models usually yield algebraic equations which are amenable to simple solution. Unsteady state models yield differential equations in time (or occasionally in position as in the case of rivers) which are soluble only in a few simple cases, the river oxygen reaeration equation being the classic example. Although numerical solution is straightforward, it is less satisfying because the sensitivity of the results to the assumed parameter values is not immediately apparent, there being no general solution. The amount of data generated by numerical solution is often overwhelming and the essential features of the chemical's behavior may be disguised in the mass of detailed data output.

Some economies are possible if equilibrium is assumed between selected compartments, an equal fugacity being assignable. This is possible if the time for equilibration is short compared to the time constant for the dominant processes of reaction or advection. For example, the rate of chemical uptake by fish from water can often be ignored (and thus need not be measured or known within limits) if the chemical has a life time of hundreds of days since the uptake time is usually only a few days. This is equivalent to the frequently used "steady state" assumption in chemical kinetics in which the differential equation for a short lived intermediate species is set to zero, thus reducing the equation to algebraic form. When the compartment contains a small amount of chemical or adjusts quickly to its environment, it can be treated algebraically.

Summary. In summary, when modeling with the fugacity concept, all equilibria can be treated by Z values (one for each compartment) and all reaction, advection and transport processes can be treated by D values. The only other quantities requiring definition are compartment volumes and emission rates or initial concentrations. A major advantage is that since all D quantities are in equivalent units they can be compared directly and the dominant processes identified. By converting diverse processes such as volatilization, sediment deposition, fish uptake and stream flow into identical units, their relative importance can be established directly and easily. Further, algebraic manipulation

is facilitated because many of the D quantities can be grouped
and error is less likely since there is no need to manipulate a
large number of symbols in diverse units.

Illustrative Application

We illustrate these concepts by applying various fugacity
models to PCB behavior in evaluative and real lake environments.
The evaluative models are similar to those presented earlier (3,
4). The real model has been developed recently to provide a
relatively simple fugacity model for real situations such as an
already contaminated lake or river, or in assessing the likely
impact of new or changed industrial emissions into aquatic environments. This model is called the Quantitative Water Air Sediment Interactive (or QWASI) fugacity model. Mathematical details
are given elsewhere (15).

The evaluative fugacity model equations and levels have been
presented earlier (1, 2, 3). The level I model gives distribution
at equilibrium of a fixed amount of chemical. Level II gives the
equilibrium distribution of a steady emission balanced by an equal
reaction (and/or advection) rate and the average residence time or
persistence. Level III gives the non-equilibrium steady state
distribution in which emissions are into specified compartments
and transfer rates between compartments may be restricted. Level
IV is essentially the same as level III except that emissions vary
with time and a set of simultaneous differential equations must be
solved numerically (instead of algebraically).

The QWASI fugacity model contains expressions for the 15
processes detailed in Figure 2. For each process, a D term is
calculated as the rate divided by the prevailing fugacity such
that the rate becomes Df as described earlier. The D terms are
then grouped and mass balance equations derived.

The following assumptions apply. The air fugacity is defined
and is not affected by the water or sediment processes. Common
fugacities apply to (i) the air, air particles and rain (f_A), (ii)
to water, suspended sediment in the lake and flowing from it (f_W),
and to the inflow water and suspended sediment f_I). If fish
concentrations are to be included, they can be calculated as
$f_W Z_B$, but the amount in fish is considered negligible.

In the most general case two differential equations are
derived, one for the water in f_W and one for the sediment in f_S.
If steady state is assumed the two equations become algebraic and
direct analytical solution is possible. An intermediate situation
can exist if the amount in the water is small compared to the
amount in the sediment, a steady state water situation can be
assumed. The water differential equation then becomes algebraic
and can be substituted into the sediment equation. Details of
these equations and their solutions are given by Mackay et al (15)

Table II gives estimated properties of a series of chlorobiphenyls and are average values for each chlorine number group of

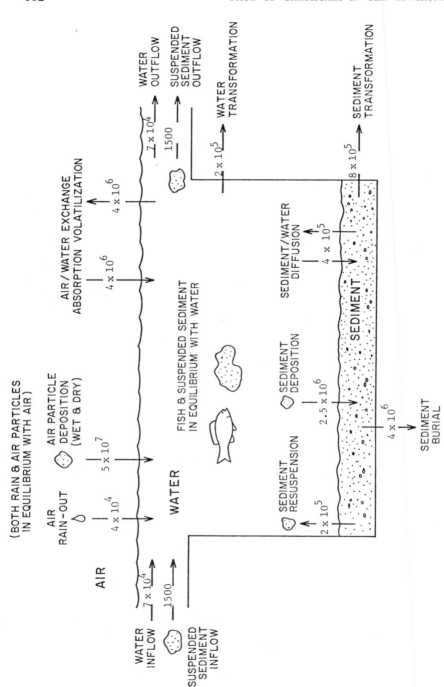

Figure 2A. Diagram of processes included in the QWASI fugacity model showing D values for a trichlorobiphenyl in a lake similar to Lake Michigan.

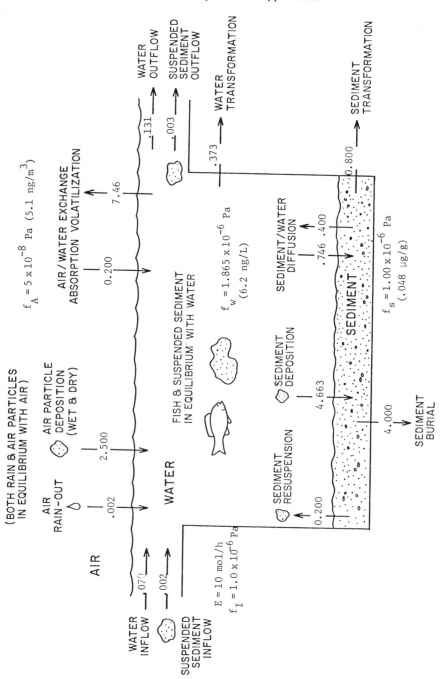

Figure 2B.   Steady state mass balance for the lake giving mass flows (mol/h), each flow being a product of D (from Figure 2A) and the appropriate fugacity.

Table II

Physical Chemical Properties of the Chlorinated Biphenyls (CBP)
Subscripts are:- A air, W water, S sediment, P suspended sediment,
B biota.

Property	Mono CBP	Di CBP	Tri CBP	Tetra CBP	Units
Mol. wt.	189	223	257	292	g/mol
H	60	60	77	76	Pa m^3/mol
Log K_{ow}	4.66	5.19	5.76	6.35	–
		Partition Coefficients			
Kp_S	757	2560	9530	37100	L/kg
Kp_P	7570	25600	95300	371000	L/kg
K_B	2290	7740	28800	112000	L/kg
		Fugacity Capacities			
Z_A	0.0004	0.0004	0.0004	0.0004	mol/m^{3}Pa
Z_W	0.017	0.016	0.013	0.013	"
Z_S	19.0	63	186	730	"
Z_P	190	630	1860	7300	"
Z_B	38.0	130	375	1500	"
		Transport Parameters			
D_{Aw}	81.2	80.6	64.1	64.6	mol/Pah
D_{ws}	16.4	16.4	13.0	13.1	"
D_{wp}	397	394	313	315	"
D_{wB}	7.60	7.72	6.24	6.24	"
		Reaction Rate Constants			
k_w	4.0×10^{-3}	2.0×10^{-4}	1.0×10^{-5}	5.0×10^{-7}	h^{-1}
k_s	0.01	5.1×10^{-4}	3.5×10^{-5}	1.13×10^{-5}	h^{-1}

congeners (13). The corresponding Z values are calculated as shown in Table II and following the general approach described earlier for evaluative environment calculations (2). These quantities should be taken as estimates rather than precise determinations since the objective here is to describe the approach and method rather than prepare a detailed evaluation or simulation model.

The evaluative lake environment is similar to the "unit world" described by Mackay and Paterson (2), consisting of a 1 km square area with an atmosphere 6000 m high, a water column 80 m deep (the approximate depth of Lake Michigan) containing suspended solids (5 parts per million by volume) and biota (considered to be fish) of 1 ppm by volume, and underlain by a sediment 3 cm deep. The bottom sediment contains 4% organic carbon and the value for suspended sediment was arbitrarily selected as ten times these bottom sediment values reflecting the enhanced sorption discussed by O'Connor and Connally (14).

Chemicals were supplied to the evaluative lake by two routes; by emissions of 0.001 mol/h directly into the water, and by advection into the air consisting of an air flow of 6.0 x 10^7 m^3/h containing 5.0 x 10^{-12} mol/m^3 (approximately 1.3 ng/m^3) resulting in a net inflow of 0.0003 mol/h. Total emissions are thus 0.0013 mol/h. This advection rate corresponds to an air residence time (volume/flowrate) of 100 hours.

Reaction rate constants are postulated as shown in Table II for degradation in water (biolysis and photolysis), in bottom sediments (probably biolysis), and for permanent burial of sediment. The values were selected from a perusal of the literature and must be regarded as speculative. A factor of 20 reduction in reaction rate constant is assumed for addition of each chlorine.

Transfer rate constants are postulated as shown in Table II, following the approach described by Mackay and Paterson (4). The air-water value selected was lower than is generally used since it appears that a low value is necessary to reconcile observed air and water concentration, and mass balances as discussed in a recent review of PCB behavior in the Great Lakes (Mackay et al. (13)).

The first set of calculations illustrates various levels of complexity for one chemical (a trichlorobiphenyl). Figure 3, 4, 5 and 6 show Level I, II, III and IV calculations, the Level IV calculations being for emissions of 0.001 mol/h for 25 years followed by response to an emission reduction by a factor of ten to 0.0001 mol/h.

The Level I calculation (Fig. 3) suggests that the dominant compartment is sediment which contains 57% of the chemical, followed by air (25%), water (10%) and suspended sediment (7%). The fish concentration is 30000 times that of the water. The absolute concentrations have no significance since they depend on the assumed amount and volumes.

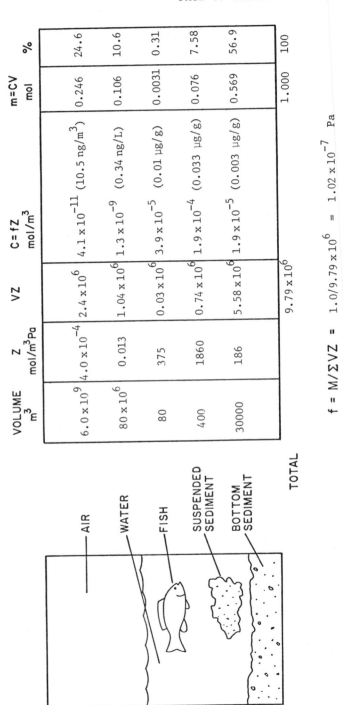

Figure 3. Level I calculation for a trichlorobiphenyl illustrating equilibrium distribution with no reaction of 1 mol of chemical.

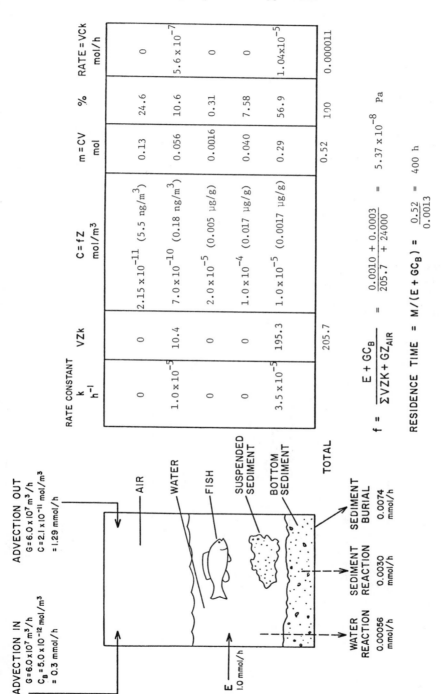

Figure 4. Level II calculation for a trichlorobiphenyl illustrating equilibrium with reaction and advection.

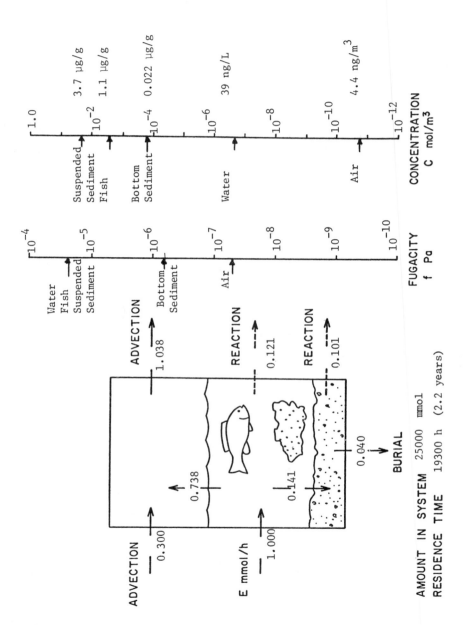

Figure 5. Level III calculation for a trichlorobiphenyl.

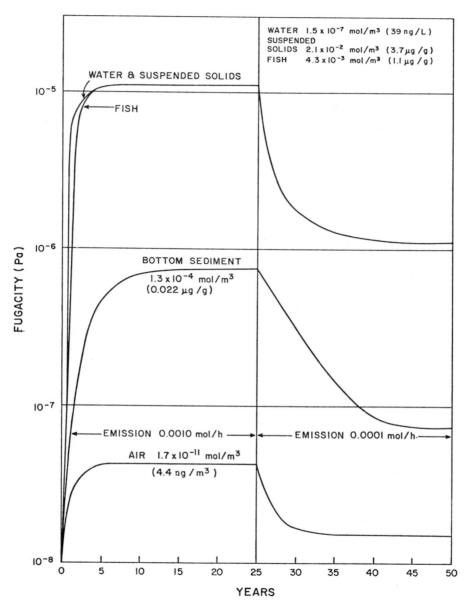

Figure 6.   Level IV calculation for a trichlorobiphenyl showing response to emission changes.

The Level II calculation (Fig. 4) has the same distribution as Fig. 3. The inflow of 1.30 m mol/h is largely removed by advection (1.284 m mol/h) with contributions by sediment burial (.0074 m mol/h), by sediment reaction (.0030 m mol/h) and by water reaction (.0006 m mol/h). This assumes that water to air transfer is rapid thus providing a resistance to this transfer, as in level III will alter the fate considerably. Atmospheric distribution of PCBs is likely to be important. The residence time of 400 h is largely controlled by air advection.

The level III calculation (Fig. 5) shows that the air-water volatilization rate constraint reduces air advective loss to 1.038 m mol/h and other reaction processes assume greater importance.

The residence time is now 2.2 years, in fair agreement with observations. The concentrations in air, water, sediment and fish are within an order of magnitude of values observed in contaminated lakes such as Lake Michigan.

The Level IV calculation (Fig. 6) shows the buildup in concentrations and fugacity to the steady state (level III values) then the subsequent decay. Clearly, sediments are slower to respond to buildup and decay, i.e. they have a longer "time constant." A tenfold drop in sediment concentration would require 15 years.

It can be concluded that these models yield a satisfactory picture of the behavior and persistence of this PCB. The dominant processes are apparent. A new chemical of similar properties is unlikely to receive environmental regulatory approval, thus the model is apparently capable of identifying such chemicals prior to their dispersal into the environment.

The Level III calculations are particularly enlightening and it is believed that they will ultimately be used for regulatory purposes. This is illustrated by Figures 7, 8, 5 and 9 which are for mono, di, tri and tetra chlorobiphenyls. The effect of increasing chlorine number is striking. The lower congeners are fairly short-lived, partition less into sediments and biota and most reaction tends to occur in the water column, advection with air and burial in sediment being relatively unimportant. As chlorine number increases, the amounts and persistence increase, more chemical partitions into the sediments and biota, while water column degradation becomes less important compared to sediment degradation and advection. Ultimately, sediment burial and advection dominate the chemical's fate. It is clear that congeners will suffer quite different environmental fates and equal emissions of each will result in very different concentrations and thus exposures. These differences should be reflected in changes in congener distribution of commercial PCB mixtures. From the hazard assessment viewpoint, it is apparent that the lower chlorine content congeners are of less concern than the longer-lived higher chlorine content congeners. This point has been made previously by Neely (11, 12).

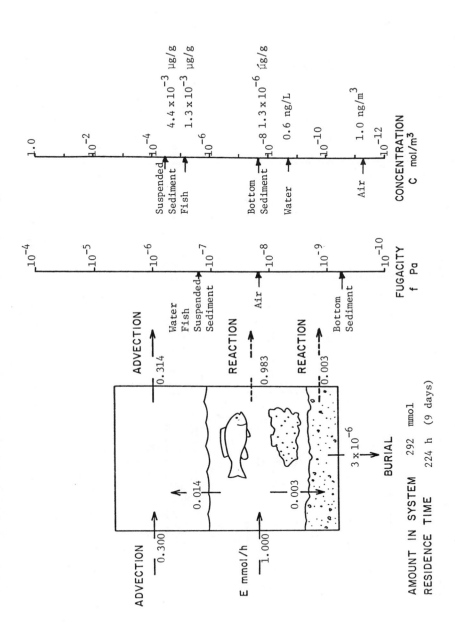

Figure 7. Level III calculation for monochlorobiphenyl.

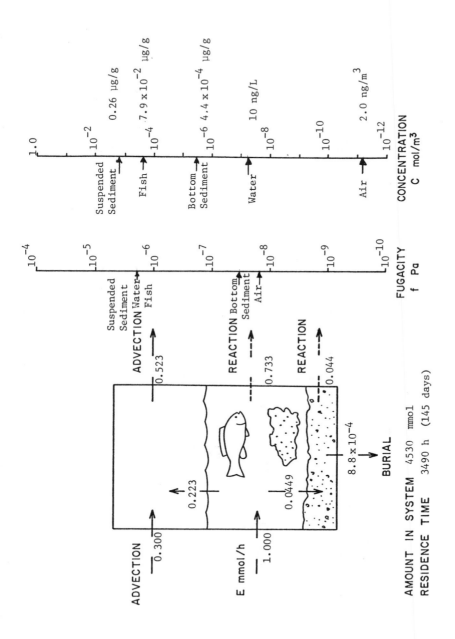

Figure 8.   Level III calculation for a dichlorobiphenyl.

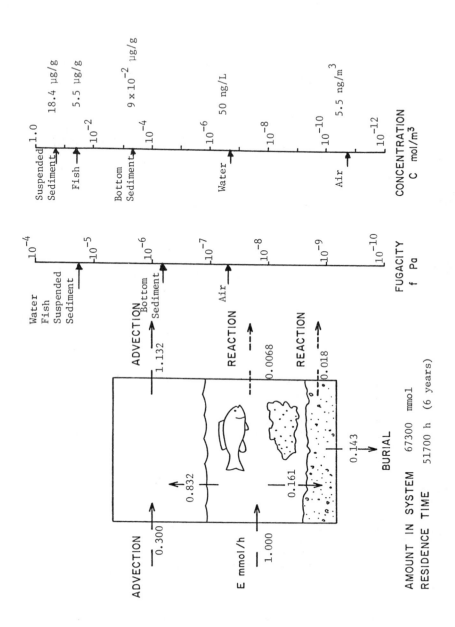

Figure 9. Level III Calculation for a tetrachlorobiphenyl.

No claim is made that PCBs behave exactly in the real environment as is suggested here, but the same principles are believed to apply.

The QWASI fugacity model was then run for a trichlorobiphenyl in a lake the size of Lake Michigan, being approximately 60,000 times the size of the evaluative environment. A detailed justification for the selection of D values is beyond our scope here but in selecting values, we have relied on recent reports by Neely (11), Rogers (15), Armstrong and Swackhamer (16), Thomann (17), and Andren (18).

To illustrate the model a steady state solution is given which would apply to the lake after prolonged steady exposure to water emission of 10 mol/h and atmospheric input from air of 5.3 ng/m³. The solution is given in Figure 2B in the form of fugacities, concentrations and transport and transformation process rates.

The dominant processes are apparently sediment deposition, sediment burial, volatilization, and deposition with air particles (i.e. dry dustfall and scavenging by rain). It is believed that the concentrations and process rates may be broadly consistent with average conditions in Lake Michigan in the early 1970s. No claim is made that the model simulates Lake Michigan precisely since the Lake has complex heterogeneous water movement and sedimentation. But the general behavior is believed to be correct and, with adjustment of the parameters, a better fit could be obtained.

If emissions were reduced by a factor of 10 to 1 mol/h and the air concentration is reduced by a factor of 5, a new steady state would emerge with a half time for change of approximately 4 years, the behavior being similar to that of the Level IV evaluative model shown earlier.

An attractive feature of the QWASI model is that the rates of the 15 processes (corresponding to 15 D values) can be compared directly and the implications of changing the assigned values can be readily explored. In some cases, the assigned values are quite speculative, partly because of uncertainty about transport rates (eg. deposition rates or volatilization mass transfer coefficients) and partly because of uncertainty about the associated equilibria (eg. Z values for depositing particles in air and water). It is possible to reach similar concentrations by adjusting sets of parameters, for example, increasing emissions and simultaneously increasing reaction rate constants. Examination of the algebraic structure of the steady state solution shows which D values act together in groups; for example, sediment burial and transformation rates can be varied individually provided that their total remains constant.

Assembling a model of a complex system such as Lake Michigan thus becomes a process of parameter selection and modification, using available process rate and equilibrium data, and resorting to intuition where necessary. It is believed, however, that this fugacity model has the ability to reproduce the reality yet remain reasonably simple. By using expressions which have a similar physical and chemical basis as those in evaluative models, the two modeling efforts can be mutually supportive and increase credibility and usage.

Progress can best be made by applying these models to new and existing chemicals at all scales, i.e. to real environments such as Lake Michigan, to rivers or small ponds, to microcosms and ultimately to laboratory flasks in which one process is isolated for study. The fugacity models described here will, it is hoped, contribute to the integration of such disparate data into more accurate profiles of chemical behavior in the environment.

Literature Cited

1. Mackay, D., Environ. Sci. & Technol. 1979, 13, 1218.
2. Mackay, D.; Paterson, S. Environ. Sci. & Technol., 1981, 15, 1006.
3. Mackay, D.; Paterson, S. "Fugacity Models for Predicting the Environmental Behavior of Chemicals", report prepared for Environment Canada 1982.
4. Mackay, D.; Paterson, S. Environ. Sci. & Technol., 1980, 16, 654A.
5. Baughman, G.; Lassiter, R. "Prediction of Environmental Pollutant Concentrations," ASTM STP 657, p 35, Philadelphia, Pa., 1978.
6. Prospectus on "Research and Development on an Exposure Analysis Modelling System (EXAMS) USEPA Athens, Ga., 1979.
7. Smith, J. H. "Environmental Pathways of Selected Chemicals in Freshwater Systems" Part I. EPA Report 600/7-77-113,1977.
8. Smith, J. H. "Environmental Pathways of Selected Chemicals in Freshwater Systems, Part II: Laboratory Studies", EPA-600/7-78-074, 1978.
9. Neely, W. B.; Mackay, D. "An Evaluative Model for Estimating Environmental Concentrations", in "Modelling the Fate of Chemicals in the Aquatic Environment", editors, Dickson, K.L. Maki, A. W. and Cairns, J., Jr., Ann Arbor Science, Ann Arbor 1982, 127-143.
10. Schmidt-Bleek, F.; Haberland, W.; Klein, A. W.; Caroli, S. Chemosphere, 1982, 11, 383.
11. Neely, W. B. "Reactivity and Environmental Persistence of PCB Isomers", in "Physical Behavior of PCBs in the Great Lakes", Mackay, D.; Paterson, S.; Eisenreich, S. J.; Simmons, M. S. (Eds.). Ann Arbor Science, 1982.
12. Neely, W. B. Chemtech, 1981, 11, 249.
13. Mackay, D.; Shiu, W. Y.; Billington, J.; Huang, G. L.; "Physical Chemical Properties of Polychlorinated Biphenyls" in Physical Behavior of PCBs in the Great Lakes", Mackay, D.; Paterson, S.; Eisenreich, S. J.; Simmons, M. S. (Eds.). Ann Arbor Science, 1982 (in press).

14. O'Connor, D. J. and Connolly, J. P. Water Res. 1980, 14, 1517.
15. Mackay, D.; Joy, M.; Paterson, S. "AQuantitative Water Air Sediment Interaction (QWASI) Fugacity Model for Describing Chemical Fate in Lakes and Rivers" submitted to Chemosphere, 1983.

RECEIVED April 15, 1983.

Environmental Fate and Transport at the Terrestrial-Atmospheric Interface

DAVID C. BOMBERGER, JULIA L. GWINN, WILLIAM R. MABEY, DANIEL TUSÉ, and TSONG WEN CHOU

SRI International, Menlo Park, CA 94025

Simple models are used to identify the dominant fate or transport path of a material near the terrestrial-atmospheric interface. The models are based on partitioning and fugacity concepts as well as first-order transformation kinetics and second-order transport kinetics. Along with a consideration of the chemical and biological transformations, this approach determines if the material is likely to volatilize rapidly, leach downward, or move up and down in the soil profile in response to precipitation and evapotranspiration. This determination can be useful for preliminary risk assessments or for choosing the appropriate more complete terrestrial and atmospheric models for a study of environmental fate. The models are illustrated using a set of pesticides with widely different behavior patterns.

Organic materials, such as pesticides or wastes placed on or near the soil surface, can undergo various fates. They can oxidize, hydrolyze, photolyze, volatilize, or biodegrade. They can be carried off the land surface into nearby streams by soil erosion or they can be leached into the soil by precipitation or irrigation. In addition, there are interactions among the various fate processes; for example, leaching into the soil retards volatilization. Also, every compound behaves differently. In assessing the overall impact that a compound might have on the environment, it is necessary to identify the important fate processes and quantify their effects and interactions. This can be done at several levels of completeness. One approach is to use simple screening estimates to determine what is expected to be the one or two dominant fate processes. On the other hand, all fate processes can be studied in great detail, and laboratory or actual field experiments may be used to gain a full understanding of what happens to a particular compound.

0097–6156/83/0225–0197$06.00/0
© 1983 American Chemical Society

The most economical route is probably to use screening studies to determine the dominant fate processes and then study only those in detail. In this paper we review some simple screening techniques that can be used to quantify volatilization and leaching rates at the soil/air interface. Volatilization and leaching rates are then compared with estimates of transformation rates to determine the compound's overall fate and identify the process requiring further study if a more exact fate assessment is required.

Method

Soil Diffusion. Water-soluble material in the soil includes material dissolved in the soil water, material dissolved in the soil air, and material adsorbed to the soil solids. The soil water-soil air equilibrium partitioning is described by Henry's law:

$$C_A = HC_w \tag{1}$$

when C_A is the gas phase concentration (mole/liter or g/cm^3) and C_w is the liquid phase concentration in the same units. H is the Henry's law constant and is unitless.

The equilibrium partitioning between soil water and soil solids is described by:

$$C_s = K_{oc} F_{oc} C_w \tag{2}$$
$$C_s = K_p C_w$$

where C_s is the concentration adsorbed on the soil (g/g or mole/g), K_{oc} is the partition coefficient on soil organic carbon, and F_{oc} is the fraction of the soil solid that is organic. Then, if we define:

V = soil volume (cm^3)

θ_A = volume fraction of soil that is air

θ_w = volume fraction of soil that is water

d = specific gravity of soil solids (g/cm^3)

ρ = soil bulk density,

the total amount of material in the soil can be expressed as:

$$
\text{Total material} = V\theta_A C_A + \frac{V\theta_w C_A}{H} + V\left(1-\theta_A-\theta_w\right)\frac{K_{oc}F_{oc}C_A}{H} \, d \qquad (3)
$$

$$
= VC_A\left(\theta_A + \frac{\theta_w}{H} + \frac{\rho K_{oc}F_{oc}}{H}\right)
$$

For convenience, we define α as:

$$
\alpha = \theta_A + \frac{\theta_w}{H} + \frac{\rho K_{oc}F_{oc}}{H} \qquad (4)
$$

The distribution between the three phases can then be described by

$$
\lambda_A = \text{Fraction compound in air phase} = \theta_A/\alpha \qquad (5)
$$

$$
\lambda_w = \text{Fraction compound in water phase} = \theta_w/H\alpha \qquad (6)
$$

$$
\lambda_s = \text{Fraction compound in solid phase} = 1 - \lambda_A - \lambda_w \qquad (7)
$$

Goring (1) showed that for many compounds, diffusion in the soil occurs primarily in only one of the three soil phases. For volatile compounds ($H > 10^{-4}$), diffusion was claimed to occur in the air phase. For nonvolatile compounds ($H < 3 \times 10^{-5}$) diffusion, if it is important, occurs in the water phase.

Volatile Compounds. For volatile chemicals, Goring (1) showed that diffusion in the soil could be described mathematically as though it occurred only in the air phase.

$$
\frac{\partial C_A}{\partial t} = D_{soil}^A \frac{\partial^2 C_A}{\partial z^2} \qquad (8)
$$

where z is the coordinate direction normal to the soil surface. D_{soil}^A was defined as:

$$
D_{soil}^A = \frac{D^A T_A}{\alpha} \qquad (9)
$$

where D^A is the diffusion coefficient of the chemical in free air and T_A is a correction for the porosity of the soil. The factor α corrects for the fact that the chemical is adsorbed on the soil solids and dissolved in the soil water. As a chemical

diffuses in the soil air, it is assumed to be replaced immediately
by material from the solid and water phases. This has the effect
of slowing the development of a concentration profile in the soil
air, which, in turn, retards the overall rate at which the
chemical is transported by diffusion. T_A is given by Hamaker (2):

$$T_A = \frac{\theta_A^{10/3}}{(\theta_A + \theta_w)^2} \qquad (10)$$

Several important assumptions are required for the
derivation:

No transport by water movement occurs.
Diffusion coefficient is a constant.
Porosity correction is constant in both time and space.
Chemicals move between the three soil phases much more
    rapidly than they diffuse in the air phase. This
    means that they appear to be in equilibrium.
Adsorption is reversible.

The last two assumptions are the most critical and are probably
violated under field conditions. Smith et al. (3) found that at
least a half-hour was required to achieve adsorption equilibrium
between a chemical in the soil water and on the soil solids.
Solution of the diffusion equation has shown that many volatile
compounds have theoretical diffusion half-lives in the soil of
several hours. Under actual field conditions, the time required
to achieve adsorption equilibrium will retard diffusion, and
diffusion half-lives in the soil will be longer than predicted.
Numerous studies have reported material bound irreversibly to
soils, which would cause apparent diffusion half-lives in the
field to be longer than predicted.

There is another underlying assumption that is also very
important—that the soil stays wet. It is well established that
organics bind much more strongly to dry soil than they do to wet
soil. K_{oc} increases in value when the soil dries out. This means
that in a dry soil ($\theta_w \approx 0$) the value of α given by Equation 4 is
too low and the estimated diffusion coefficient is too large.
Spencer and Cliath (4) measured very dramatic increases in
sorption of lindane as soil moisture was decreased below the
amount required for monolayer coverage. Ehlers et al. (5) showed
decreases in the effective soil diffusion coefficient as the soil
dried. Because under field conditions the soil can dry out,
predicted soil diffusion half-lives will almost always be shorter
than those that actually occur.

Farmer (6) reviewed the various diffusion models for soil and developed solutions for several of these models. An appropriate model for field studies is a nonsteady state model that assumes that material is mixed into the soil to a depth L and then allowed to diffuse both to the surface and more deeply into the soil. Material diffusing to the surface is immediately removed by diffusion and convection in the air above the soil. The effect of this assumption is to make the concentration of a diffusing compound zero at the soil surface. With these boundary conditions the solution to Equation 8 can be converted to the useful form:

$$f = \sqrt{\frac{4D^A_{soil}t}{\pi L^2}} \left[1 - \exp\left(\frac{-L^2}{4D^A_{soil}t}\right) \right] + \mathrm{erfc} \sqrt{\frac{L^2}{4D^A_{soil}t}} \qquad (11)$$

where f is the fraction of the material originally in the soil that remains at any time t. It can be shown that f = 0.5 when $L^2/4D^A_{soil}t = 1.04$, which means that the half-life for diffusion is $L^2/4.16D^A_{soil}$.

Nonvolatile Compounds. The same formal development can be used to develop diffusion equations for nonvolatile compounds. The result is:

$$\frac{\partial C_w}{\partial t} = D^W_{soil} \frac{\partial^2 C_w}{\partial^2 z} \qquad (12)$$

$$D^W_{soil} = \frac{D^W T_w}{H \alpha} \qquad (13)$$

where D^W is the compound's diffusion coefficient in free water and T_w is a correction factor for the soil porosity.

Diffusion coefficients in water are much smaller than diffusion coefficients in air. As an example, for oxygen, $D^A = 1.75 \times 10^{-1}$ cm^2/s but $D^W = 2.1 \times 10^{-5}$ cm^2/s. Consequently, the predicted soil diffusion half-lives for volatile compounds range from hours to days, whereas the predicted half-lives for nonvolatile compounds range from weeks to months. This long diffusion half-life for nonvolatile compounds results in a violation of one of the derivation assumptions: that no transport by water movement occurs. Over several weeks, a significant fraction of soil water will evaporate, and the movement of water through the pores becomes the principal transport mechanism for dissolved nonvolatile organics. The water evaporation rate determines the compound's mass transport rate, and overall volatilization rates will be slow.

Mass Transport and Leaching. Organic substances can be moved through the soil by soil water flowing downward as a result of infiltration induced by precipitation and irrigation and upward as a result of evapotranspiration. The rate of movement will be affected by the adsorption equilibrium of the substance between soil water and soil solids. In general, substances that are strongly adsorbed move much more slowly than the soil water, and those that are weakly adsorbed may move at the same rate. Several mathematical models have been proposed for describing this movement. The earlier models were developed from the study of chromatography (7, 8, 9). These models assumed a pointwise equilibrium adsorption throughout the soil profile. Davidson et al. (10) also assumed this pointwise equilibrium. Oddson et al. (11) and Lindstrom et al. (12) used a kinetic adsorption model. All these models could be solved analytically, but required steady-state soil water conditions. These models are of limited use because in actual field situations soil water content and movement change with time.

Davidson et al. (13) developed numerical solutions of the differential equation for solute transport for a model that included both the transient and steady-state soil water conditions after precipitation events. The model also included the effects of diffusion and dispersion. The model did not, however, include the effects of evapotranspiration. Evapotranspiration causes a movement of water to the soil surface that carries organic material with it. This water movement is responsible for the well known "wick effect," (14) and not including it severely limits the utility of the model.

Because the more complicated model that required numerical solution still neglected important effects, we chose to use a simple analytical model for convenience. We chose Oddson's because its major features had been verified by Huggenberger (15, 16) for lindane, one of the compounds in our study. Oddson included the kinetics of adsorption by assuming that the rate of adsorption is proportional to the difference between the amount that has already adsorbed and the equilibrium value:

$$\frac{\partial C_s}{\partial t} = \beta(K_p C_w - C_s) \qquad (14)$$

β (1/h) is a constant describing the rate at which adsorption equilibrium is achieved.

The overall transport of chemical by water movement is described by:

$$v\frac{\partial C_w}{\partial x} + \theta_w \frac{\partial C_w}{\partial t} + \rho \frac{\partial C_s}{\partial t} = 0 \qquad (15)$$

where v is the superficial water velocity in the soil. [The superficial velocity is equal to the rate at which water is applied to the soil surface cm^3/cm^2-s .] This formulation does not include the effects of diffusion or dispersion, which makes the solutions tractable.

Oddson solved an initial value problem that described the convective movement of an organic down from the soil surface for the following specific conditions:

No organic chemical is present in the soils originally.

The organic chemical is placed on the soil surface and carried into the soil by water applied to the soil surface. Assuming the resulting concentration at the surface is constant at C_{sat} (compound solubility) for the time T that it takes to dissolve all the material and is zero thereafter, we have

$$C(0,t) = C_{sat} \text{ for } 0 < t < T$$

and

$$C(0,t) = 0 \text{ for } t > T$$

The actual solutions that were developed are not shown here because they are not needed for a screening analysis. The major features of the solutions include the following:

K_p influences the depth of maximum concentration of organic in solution, but does not affect the value of that concentration. The organic chemical in solution will move through the soil as a gaussian peak. The lower K_p, the more spread out the peak will be. The depth of movement of maximum concentration is equal to the depth of water penetration divided by $\rho K_p/\theta_w$. For field studies, an appropriate value of θ_w to use is the field capacity, which is the water content that develops in a soil that is saturated and then allowed to drain freely.

The concentration of material adsorbed on the soil also moves down as a peak. The position of maximum adsorbed material is about the same as for the maximum concentration in solution.

These two conclusions are the important results from the model because they enable us to say how deep into the soil profile the majority of an organic chemical penetrates due to water inputs. If V centimeters of water are applied to a soil surface, then the water penetrates the soil to a depth of V/θ_w . If V is sufficient to dissolve all the organic chemical present, then depth where the maximum concentration of chemical in the soil will be found is $V/\rho Kp$.

The rate of adsorption, which is described by the parameter β , affects the shape of the concentration peak. Large values of β cause sharp peaks, whereas small values cause wide peaks. Various values have been found for β (11). They range from 0.025 to 5.0, with the lower values that imply a slow approach to equilibrium associated with soils high in organic content. It appears also that β may not really be a constant, being larger when adsorption is far from equilibrium and smaller later. For soils with the smaller values of β , some organic material may be found at the depth of water penetration even though the bulk of the organic may be much nearer to the surface.

Soil. The fate of organics is highly dependent on the soil properties. The models for diffusion and mass transport show that highly organic soils retard diffusion and mass transport by strengthening the sorptive attraction of the soil particles for organics. Highly porous and dry soils (θ_A large) foster diffusion because they offer adequate air space for diffusion. For this investigation we chose a soil with "typical properties" so that the behavior of different compounds could be compared. The soil has a porosity of 50%, the soil particles themselves have a density of 2.5 g/cm^3 (a clay), and the bulk density of the dry soil is 1.25 g/cm^3. The soil field capacity is 30% and its organic content is 1%. Soil organic contents can vary from almost zero for sandy soils to 11% or more for organic mucks. Two percent is considered highly organic, and 1% is fairly representative of agricultural soil.

Diffusion Coefficients. Diffusion coefficients in air were estimated using the Fuller, Schetller, and Giddings correlation (17) even though for some compound measured values were available.

Adsorption Partition Coefficients. Experimental K_{oc} values were used when available; otherwise, the K_{oc} values were estimted. We used a correlation between aqueous solubility (C_{sat}) and K_{oc} that contained data for pesticides and a group of polar and nonpolar organic chemicals collected by Kenega and Goring (18) and Smith et al. (3):

$$\log K_{oc} = -0.27 - 0.782 \log C_{sat} \qquad (16)$$

This relationship does not include the correction recommended by
Yalkowsky (19) for including differences between solid and liquid
compounds because Yalkowsky's work was not available at the
beginning of the study. Yalkowsky showed that solids and liquids
do not fit well into the same correlation unless the solid
solubilities are corrected for the entropy of melting. The error
introduced by not including the correction is not significant for
this screening analysis.

<u>Henry's Law Constants</u>. When available, experimental values of
Henry's law constants were used. When experimental values could
not be found, values were estimated using the method outlined by
Mackay (20, 21, 22):

$$H = RTP/C_{sat} \qquad (17)$$

where R is the universal gas constant, T is the absolute
temperature, P is the vapor pressure of the compound in its stable
form at 20°C, and C_{sat} is its aqueous solubility at 20°C. For
solid compounds where vapor pressures were only available for
melts at elevated temperature, the vapor pressure of the subcooled
liquid at 20°C was estimated by using the Clausius-Clapeyron
equation. The vapor pressure of the subcooled liquid was not
corrected to give the vapor pressure of the stable solid because
Yalkowsky's (23) work on heats of fusion was not available at the
beginning of the study. The error introduced by not including the
correction is not significant for this screening analysis.

<u>Transformation Rates</u>. A literature search was conducted to
determine rates of oxidation, hydrolysis, photolysis, and
biodegradation. When no values were found, we made estimates
based on our experience, known rates for similar compounds, and
structure-activity relationships. In cases where there was great
uncertainty, a transformation rate of zero was assumed so that the
compound would be considered persistent. This would force a more
detailed fate assessment to be conducted if considerations of
toxicity indicated that the compound might be hazardous.

<u>Results and Discussion</u>

Results of volatilization and leaching estimations are
reported for six pesticides that span a wide range of the
physical/chemical properties that affect fate at the soil/air
interface. The pesticides are Mirex, toxaphene, methoxychlor,
lindane, malathion, and dibromochloropropane (DBCP). These
particular pesticides were chosen for discussion here because they
illustrate the methods for assessing the fate of organics at the

soil/air interface. They were not chosen out of any concern for their toxicity or long-term environmental effects. However, several of the pesticides are considered priority pollutants by the EPA. Because all but malathion are currently restricted or no longer registered for use, our discussion is largely academic.

Table I shows the results of calculating a soil diffusion coefficient and soil diffusion half-lives for the pesticides. The 10% moisture level specified means that the soil is relatively dry and that 40% of the soil volume is air available for diffusion. Complete calculations were not made for methoxychlor, lindane, and malathion because, based on Goring's criteria for the Henry's law constant, they are not volatile enough to diffuse significantly in the gas phase. This lack of volatility is reflected in their low values of λ_A . These materials would move upward in the soil only if carried by water that was moving upward to replace the water lost through evapotranspiration at the surface. Mirex has a very high Henry's law constant. On the basis of Goring's criteria, Mirex should diffuse in the soil air; but, because of its strong adsorption, it has a very large α and consequently a very small soil air diffusion coefficient. The behavior of Mirex shows that Goring's criteria must be applied carefully.

Half-lives for Mirex, toxaphene, and DBCP were calculated by assuming that they were mixed into the top 10 cm of the soil so that the effects of Henry's law constants and sorption partition coefficients could be compared on a common basis for all chemicals. For DBCP, this mixing depth partially reflects the method of use because after application it was flushed into the soil profile by flooding or irrigating the fields. For toxaphene, the mixing depth reflects the fact that in some soils it is persistent from one year to the next and would be mixed into the soil profile by plowing. For Mirex, the 10-cm mixing depth is probably totally artificial because Mirex is not plowed into the soil; also, as the discussion on leaching shows it is not mobile in the soil.

The pesticides are arranged in order of decreasing adsorption, and although the adsorption coefficient changes by six orders of magnitude, the change in diffusion half-life is not directly correlated with this change. The Mirex, which is very strongly adsorbed and only moderately volatile, hardly diffuses at all. It stays where it is placed. The toxaphene, which is strongly sorbed but is also quite volatile, is predicted to have a soil diffusion half-life of only 9 days. The DBCP, which is much less strongly adsorbed, has a diffusion half-life of only 1.2 days even though it is less volatile than toxaphene.

In Table II, we use DBCP to illustrate how soil properties and conditions can affect diffusion half-lives. Increasing the

Table I. Diffusion Behavior of Pesticides in a Soil with 10% Moisture

Pesticide	K_{oc}	H	λ_s	λ_A	D^A_{soil} (cm^2/s)	$t_{1/2}$ (day)
Mirex	2.4×10^7 [a]	0.03 [b]	~ 1.0	4×10^{-8}	7.5×10^{-10}	0.3×10^6
Toxaphene	2.1×10^5 [c]	9.0 [b]	0.998	1.4×10^{-3}	3.0×10^{-5}	9
Methoxychlor	8.0×10^4 [d]	3.0×10^{-5} [b]	~ 1.0	1×10^{-10}	$--$ [e]	$--$
Lindane	1.3×10^3 [f]	5.0×10^{-5} [g]	0.994	1.2×10^{-6}	$--$ [e]	$--$
Malathion	2.3×10^2 [a]	4.9×10^{-6} [b]	0.967	6.6×10^{-7}	$--$ [e]	$--$
DBCP	40 [a]	0.01 [b]	0.828	6.7×10^{-3}	2.3×10^{-4}	1.2

[a] Smith et al., (3).

[b] Estimate based on vapor pressure and solubility.

[c] Mulkey (24).

[d] Karickhoff et al (25).

[e] D^A_{soil} is not shown because material cannot diffuse significantly in soil air because of low volatility.

[f] Hamaker and Thompson (26).

[g] Hamaker (2).

organic content of the soil increases adsorption and increases α, which, in turn, decreases the diffusion coefficient in the soil. An even more dramatic effect is obtained by increasing soil moisture. This fills the pore spaces and blocks diffusion in the gas phase by decreasing T_A. Changing soil moisture from 10% to 25% increases soil half-lives by a factor of six. This effect was discussed by Farmer et al. (27) in a study where they noted that keeping soil cover over a dump site wet was one means of controlling diffusion of hexachlorobenzene from the dump into the atmosphere.

Table II. Volatilization Half-Lives of DBCP Mixed Into
10 cm of Several Kinds of Soils
(day)

| | Soil Condition | |
Soil Organic Content F_{oc}	Dry Soil (10% water, 40% air)	Wet Soil (25% water, 25% air)
0.0025	0.6	3.6
0.01	1.2	7.2
0.05	5.2	26.2

Table III illustrates the impact of adsorption on the leaching of organic chemicals in the soil. A water input of 305 cm was used, which is equivalent to a full year of precipitation in the eastern United States. In a soil with a field capacity of 30%, the water would penetrate 1017 cm. Mirex with a very large K_{oc} is practically immobile; after a full year of precipitation, it is still on the surface. It is likely that any compound adsorbed this strongly would be carried off the land surface by soil erosion instead of being leached into the soil. In contrast, DBCP, which is very weakly adsorbed, penetrates the soil profile almost as far as the water does.

Table III. Leaching Behavior of Pesticides in a Typical
Soil with a Field Capacity of 30%

Pesticide	Adsorption Coefficient K_p	Depth of Maximum Concentration After 305 cm of Water Applied (cm)
Mirex	2.4×10^5	0.001
Toxaphene	2.1×10^3	0.1
Methoxychlor	8.0×10^2	0.3
Lindane	1.3×10	19
Malathion	2.3	102
DBCP	4×10^{-1}	610

For each of the model compounds, some material will have leached deeper into the soil than is shown in the table. The model calculates only the position of maximum concentration. For a compound like DBCP, which has a very weak adsorption interaction with the soil, the concentration profile will be spread out. DBCP would probably be found at low concentrations at the 1017 cm level. For the strongly adsorbed compounds, such as toxaphene and methoxychlor, the concentration peak will be narrow, and the depth of maximum concentration is the depth where most of the material is.

Table IV uses DBCP to illustrate the effect of soil organic carbon concentration on leaching mobility. In the very low organic carbon soil, there is almost no adsorption for the compound and the leaching model breaks down because it predicts penetration depths greater than the water penetration. In these cases the prediction is adjusted to show compound and water penetration depths as the same.

Table IV. Effect of Soil Organic Carbon Levels
on Leaching Depth
(cm)

Soil Organic Content (F_{oc})	Total Precipitation Water	
	25 cm	305 cm
0.0025	83[a]	1017[a]
0.01	50	610
0.05	10	122

[a]There is so little adsorption that the material moves with the water front. The mathematical model breaks down under these circumstances.

Volatilization and leaching interact with each other and other fate processes. Two of the pesticides, DBCP and lindane, are discussed in some detail to illustrate some of the interactions. The other four are discussed only briefly.

DBCP. The predictions suggest that DBCP is volatile and diffuses rapidly into the atmosphere and that it is also readily leached into the soil profile. In the model soil, its volatilization half-life was only 1.2 days when it was assumed to be evenly distributed into the top 10 cm of soil. However, DBCP could be leached as much as 50 cm deep by only 25 cm of water, and at this depth diffusion to the surface would be slow. From the literature study of transformation processes, we found no clear evidence for rapid oxidation or hydrolysis. Photolysis would not occur below the soil surface. No useable data for estimating biodegradation rates were found although Castro and Belser (28) showed that biodegradation did occur. The rate was assumed to be slow because all halogenated hydrocarbons degrade slowly. DBCP was therefore assumed to be persistent.

The predicted fate of DBCP depends on the amount of water that enters the soil after it has been applied. With very little water and leaching depths of 10 cm or less, volatilization will dominate and most DBCP will leave the soil. On the other hand, if enough water enters the soil to leach DBCP below 10 cm, volatilization will be retarded, and DBCP could remain in the soil profile for an extended period. In either case, however, because DBCP does not adsorb strongly to soil, the concentration profile from leaching is quite spread out. Even if the bulk of the DBCP is leached only 10 cm deep, some material would be leached much deeper where diffusion to the surface would be slow. Contamination of shallow ground waters might be possible and should be considered in a more detailed fate assessment. The detailed assessment should obtain useful values for oxidation, hydrolysis, and biodegradation rates.

Lindane. The volatilization calculations predicted that lindane that has entered the soil profile does not volatilize by diffusion but rather by mass transport due to water evaporation at the soil surface. On the other hand, the leaching calculations show that lindane is not highly mobile in the soil profile: 305 cm of water move it only 10 cm into the soil profile from the surface. There is no evidence for oxidation, and because aquatic hydrolysis half-lives are at least six months (29), hydrolysis rates in the soil will be slow. Photolysis is not expected, but biodegradation does occur. A major biodegradation product is γ-pentachloro-cyclohexane (PCCH), which is much more volatile than lindane (30). The results of studies of the biodegradation rates scatter widely, and reported half-lives range from weeks to months. If lindane were applied to the surface in a climate where precipitation exceeds evapotranspiration, it could be leached deep into the soil over time and could contaminate ground water if the slower biodegradation rates applied. In a situation where evapotranspiration and precipitation or irrigation were in balance, lindane could move up and down in the soil profile with little net transport in either direction. The major loss would be by bioconversion to PCCH and then volatilization of the PCCH. Actual field experiments (31) have confirmed this behavior.

Mirex. Mirex does not leach into the soil profile and is predicted to volatilize only slowly. There is no evidence for any rapid transformation so it should be considered persistent. Because it is so strongly adsorbed to the soil and stays on the surface, a major loss from terrestrial systems would probably be erosion and transport into surface waters.

Toxaphene. Toxaphene is apparently strongly adsorbed and should not move in the soil profile. Because of its strong sorptive interaction with soils, some material may erode into surface waters during irrigation or precipitation events. There is no evidence for oxidation, hydrolysis, or biodegradation (29). Photolysis probably would not occur. However, volatilization is predicted to be rapid. Our predictions of rapid volatilization are consistent with measurements made by Sieber et al. (32). Normally, toxaphene would be considered nonpersistent in the soil environment because of its volatility. However, field studies show that toxaphene is persistent. The difference between prediction and reality probably results because toxaphene is not a single compound but a mixture of hundreds of components. Although the overall mixture is strongly adsorbed and not mobile in the soil, some individual components may be mobile and leach deep into the soil where diffusion to the surface would be slow. In addition, although the mixture has a high volatility, some components may be relatively nonvolatile and persist in the soil. Thus, mixtures may not be suitable candidates for the screening approach developed here.

Methoxychlor. Methoxychlor is strongly adsorbed to the soil and does not leach, and volatilization is slow. There is no evidence for oxidation, and although photolysis is rapid in aquatic systems, it is assumed not to occur in the soil environment. The hydrolysis half-life is a year in aquatic systems (33) and probably longer in soil systems because of adsorption. Biodegradation does occur in soil systems, however, with a half-life of from 1 to 3 weeks (34). Methoxychlor would not persist in the soil environment.

Malathion. Malathion is not strongly adsorbed and could leach deeply into the soil. It is not volatile and would volatilize from the soil only as rapidly as it was carried to the surface by evapotranspiration. Biotransformation is rapid (35), however, so it should not persist in the soil environment.

Summary

A screening analysis has been developed that can be conducted quickly with only a modest investment of time and money. It is useful in determining both how long a compound might persist near the soil surface and where it goes if it does not persist. If a detailed fate assessment is required, the screening helps to determine whether complicated leaching or runoff models will be needed. The analysis has been applied here to pesticides but there are situations where it can be applied to other organics. Some aspects of migration of organics from landfills and land disposal sites can be considered. It can also be applied to determine the fate of organic ingredients in some consumer products where, if it were coupled to considerations of toxicity and environmental impact, it would be a useful tool for choosing between competing ingredients.

Acknowledgment

This work was conducted under Contract 68-01-3867 with the United States Environmental Protection Agency.

Literature Cited

1. Goring, C.A.I.   "Agricultural Chemicals in the Environment: A Quantitative Viewpoint."   Chapter 13 in "Organic Chemicals in the Soil Environment"; Marcel Dekker: New York, 1972.
2. Hamaker, J. W.   "Diffusion and Volatilization."   Chapter 5 in "Organic Chemicals in the Soil Environment"; Marcel Dekker: New York, 1972.
3. Smith, J. H.; Mabey, W. R.; Bohonos, N.; Holt, B. R.; Lee, S. S.; Chou, T.-W.; Bomberger, D. C.; Mill, T.   "Environmental Pathways of Selected Chemicals in Freshwater Systems, Part II: Laboratory Studies;" U. S. Environmental Protection Agency, Office of Research and Development, Southeast Environmental Research Laboratory: Athens, Georgia, 1978; EPA 600/7-78-074.
4. Spencer, W. F.; Cliath, M. M.   Soil Sci. Soc. Amer. Proc. 1970, 34, 574.
5. Ehlers, W.; Farmer, W. J.; Spencer, W. F.; Letey, J.   Soil Sci. Soc. Amer. Proc. 1969, 33, 567.
6. Farmer, W. J.; Letey, J.   "Volatilization Losses of Pesticides from Soils"; U. S. Environmental Protection Agency, Office of Research and Development, Southeast Environmental Research Laboratory: Athens, Georgia, 1974; EPA 660/2-74-054.
7. Lapidus, L.; Amundsen, N. R.   J. Phys. Chem. 1959, 56, 984.
8. Hashimoto, I., Deshpande, K. B.; Thomas, H. C.   Ind. Eng. Chem. Fundam. 1964, 3, 213.
9. King, P. H.; McCarty, P. L.   Soil Science 1968, 106, 248.
10. Davidson, J. M.; Rieck, C. E.; Santelmann, P. W.   Soil Sci. Soc. Amer. Proc. 1968, 32, 629.
11. Oddson, J. K.; Letey, J.; Weeks, L. V.   Soil Sci. Soc. Amer. Proc. 1970, 34, 412.
12. Lindstrom, F. T.; Boersma, L.; Stockard, D.   Soil Science 1971, 112, 291.
13. Davidson, J. M.; Brusewitz, G. H.; Baker, D. R.; Wood, A. L.   "Use of Soil Parameters for Describing Pesticide Movement Through Soils," U. S. Environmental Protection Agency, Office of Research and Development: 1975; EPA 660/2-75-009.
14. Spencer, W. F.; Cliath, M. M.   J. Environmental Quality 1973, 2, 284.
15. Huggenberger, F.   "Adsorption and Mobility of Pesticides in Soils." Ph. D. Thesis: University of California, Riverside, 1972.
16. Huggenberger, F.; Letey, J.; Farmer, W. J.   Soil Sci. Soc. Amer. Proc. 1972, 36, 544.

17. Sherwood, T. K.; Pigford, R. L.; Wilke, C. R.   "Mass
     Transfer"; McGraw Hill, Inc.: San Francisco, CA, 1975.
18. Kenaga, E. E.; Goring, C.A.I.   "Relationship Between Water
     Solubility, Soil Sorption, Octanol-Water Partioning, and
     Bioconcentration of Chemicals in Biota," ASTM, Third Aquatic
     Toxicology Symposium: October 17-18, 1978.
19. Yalkowsky, S. H.; Valvani, S. C.   J. Pharm. Sci. 1980, 69,
     912.
20. Mackey, D.; Wolkoff, A. W.   Environ. Sci. Technol. 1973, 7,
     611.
21. Mackay, D.; Leinonen, P. J.   Environ. Sci. Technol. 1975, 9,
     1178.
22. Mackay, D.; Shiu, W. Y.; Sutherland, R. P.   Environ. Sci.
     Technol. 1979, 13, 333.
23. Yalkowsky, S. H.   Ind. Eng. Chem. Fundam. 1979, 18, 108.
24. Mulkey, L.   U. S. Environmental Protection Agency: Athens,
     GA; personal communication.
25. Karickhoff, S. W.; Brown, D. S.; Scott, T. A.   Water Research
     1979, 13, 241.
26. Hamaker, J. W.; Thompson, J. M.   "Adsorption" Chapter 2 in
     "Organic Chemicals in the Soil Environment"; Marcel Dekker:
     New York, 1972.
27. Farmer, W. J.; Yang, M.-S.; Letey, J.; Spencer, W. F.   "Land
     Disposal of Hexachlorobenzene Wastes; Controlling Vapor
     Movement in Soil"; U. S. Environmental Protection Agency,
     Office of Research and Development, Municipal Environmental
     Research Laboratory: Cincinnati, Ohio, 1980; EPA 600/2-80-
     119.
28. Castro, C. E.; Belser, N. O.   Environ. Sci. Technol. 1968, 2,
     779.
29. Callahan, M. A.; Slimack, M. W.; Gabel, N. W.; May, I. P.;
     Fowler, C. F.; Freed, J. R.; Jennings, P.; Durfee, R. L.;
     Whitmore, F. C.; Maestri, B.; Mabey, W. R.; Holt, B. R.;
     Gould, C.   "Water Related Fate of 129 Priority Pollutants,
     Volume I"; U.S. Environmental Protection Agency, Office of
     Water Planning and Standards: Washington, DC, 1979; EPA
     440/4-79-029a.
30. Cliath, M. M.; Spencer, W. F.   Environ. Sci. Technol. 1972,
     6, 910.
31. Cliath, M. M.; Spencer, W. F.   Soil Sci. Soc. Amer. Proc.
     1971, 35, 791.
32. Sieber, J. N.; Madden, S. C.; McChesney, M. M.; Winterlin, W.
     L.   J. Agricultural and Food Chemistry 1979, 27, 284.
33. Wolfe, N. L.; Zepp, R. G.; Paris, D. F.; Baughman, G. L.;
     Hollis, R. C.   Environ. Sci. Technol. 1977, 11, 1077.
34. Castro, T. F.; Yoshida, T.   J. Agricultural and Food
     Chemistry 1971, 19, 1168.
35. Klein, S. A.; Jenkins, D.; Wagenet, R. J.; Biggar, J. W.;
     Yang, M.   "An Evaluation of the Accumulation, Translocation,
     and Degradation of Pesticides at Land Wastewater Disposal.

RECEIVED May 2, 1983.

Interactions Between Dissolved Humic and Fulvic Acids and Pollutants in Aquatic Environments

CHARLES W. CARTER [1] and I. H. SUFFET

Drexel University, Environmental Studies Institute, Philadelphia, PA 19104

We have made quantitative measurements of the binding of organic compounds to dissolved humic and fulvic acids. The extent of binding increases as a compounds octanol/water partition coefficient increases or as its water solubility decreases. Humic acids bind compounds to a greater extend than fulvic acids, but there are large differences between different humic and fulvic acids.

A number of mathematical models have been developed in recent years which attempt to predict the behavior of organic water pollutants.[1,2,3] Models assume that compounds will partition into various compartments in the environment such as air, water, biota, suspended solids and sediment. The input to the models includes the affinity of the compound for each of the compartments, the rate of transfer between the compartments, and the rates of various degradation processes in the various compartments. There is a growing body of data, however, which indicates that the models to date may have overlooked a small but significant interaction. A number of authors have suggested that a portion of the compounds in the aqueous phase may be bound to dissolved humic materials and are not therefore truly dissolved.

If this binding does occur, then one would expect very strongly bound compounds to show an unusual affinity for the aqueous phase. This could increase the mobility of these compounds in the environment. It is likely that the bound fraction will undergo phase transfers and degradation at different rates than the free truly dissolved fraction of a dissolved pollutant. If this is the case, then an observed equilibrium between a pollutant in the free and bound states could significantly affect its environmental behavior.

[1] Current address: Versar Inc., 6850 Versar Center, Springfield, VA 22151

A number of different authors have made quantitative
measurements of the extent of binding of organic compounds to
dissolved humic materials. Wershaw et al[4] and Porrier et
al[5] have measured the binding of DDT to humic materials from
different origins. More recently, Carter and Suffet[6] and
Carter[7] have measured the binding of a variety of compounds
including pesticides, polynuclear aromatics and phthalates to
various humic materials. Means and Wiyajaratne[8] have
quantitatively measured the binding of Atrazine and Linuron to
colloidal estuarine organic matter. Diachenko[9] has measured
the binding of a number of chlorinated organic compounds to
various humic materials. All of these authors indicate that a
substantial fraction of a pollutant found in the aqueous phase
may in fact be bound to dissolved humic materials, and they
suggest that this may change the behavior of the compound.

There is more current literature which speaks to exactly
this point. Hassett[10] has found that the sorption of organic
compounds from water to suspended particles is altered in the
presence of dissolved organic carbon. Griffin Chian[11] and
Diachenko[9] have found that the volatility of organic compounds
in water decreases when humic materials are present. Perdue[12]
has found that the rate of hydrolysis of the octyl ester of
2,4-D is decreased in the presence of humic materials.
Landrum[13] and Hassett[14] found that dissolved organic matter
can affect the analysis of organic compounds. Eisenreich[15]
has presented data on the partitioning of PCB's between solids
and pore water in Lake Superior sediments. His data suggest
that the dissolved organic carbon in the pore water may
significantly affect the mobility of the PCB's. Zepp et al[16]
have reported that the rate of photolysis of some organic
compounds increases in the presence of humic materials.
Leversee[17] has found that the extent of bioaccumulation of
some polynuclear aromatic compounds changes in the presence of
humic materials.

From the literature reviewed above it is clear that a number
of authors have determined that certain compounds can and do
bind to dissolved humic materials. Other authors have invoked
this binding phenomenon to explain otherwise peculiar data. It
would be desirable to incorporate this binding into
environmental fate models, but there is not much data on the
phenomena and there are few methods available to collect more of
this data.

This paper will summarize some of our work on the binding of
pesticides and other pollutants by dissolved humic materials.
The methods used in our work and in the work of other authors
will also be presented. Hopefully this will stimulate more
widespread interest in understanding these phenomena and will
facilitate the collection of meaningful data.

Methods Available for Quantitatively Measuring the Extent of Binding of Organic Pollutants to Dissolved Humic Materials

Three methods were used in this research to measure the extent of binding of organic pollutants to dissolved humic materials. They were equilibrium dialysis, solubility measurements and changes in sorption behavior in the presence of humic materials. Other authors have used solubility measurements, ultrafiltration and volatilization measurements. The methods will be described in the following paragraphs.

In a dialysis experiment, a dialysis bag containing the dissolved humic materials is placed in a solution of a pollutant (preferably radiolabeled). The dialysis tubing is chosen so the pollutant is free to diffuse through the bag while the humic materials are retained inside the bag. The solution is shaken at constant temperature until it comes to an equilibrium point. At equilibrium, the pollutant inside the dialysis bag consists of two fractions: that truly dissolved and the bound to the humic materials. The concentration of pollutant on the outside of the dialysis bag consists only of the free, truly dissolved fraction. Any increase of the pollutant concentration on the inside of the dialysis bag is due to binding by dissolved humic materials. A series of dialysis experiments, therefore, can measure the bound fraction concentration as a function of the free concentration.

In a solubility experiment the solubility of the compound of interest is measured in the presence and absence of dissolved humic materials. Two techniques were used to measure solubility: a shake and filter method similar to that used by Yalkowsky, and a flow through column technique similar to that used by May et al.[19] The measured solubilities of a number of compounds in our experiments were always higher in the presence of humic materials. This increase in the solubility is due to the binding of the compound by humic materials. In the presence of humic materials the measured solubility consists of two fractions; free and bound. The free concentration should be the same in the presence or absence of humic materials. The difference between the solubilities of the compound in the presence and absence of humic materials is therefore a measurement of the bound fraction.

The third technique we used was a measurement of changes in the sorption behavior of a compound in the presence of humic materials. A thin film of OV-1, a methyl silicone gum used as a chromatographic stationary phase, was plated on the bottom of a 60 ml Hypo-Vial (Pierce Chemical Co., Rockford, Il.). A solution of radiolabeled pollutant was added to the vial in either buffered distilled water in a solution of humic materials. Again, it is assumed that the pollutant is solution consists of two fraction; free and bound. It is also assumed that the bound fraction will not sorb to the film of OV-1. This is a fairly reasonable assumption since the humic polymer is negatively charged at the pH values studied and will have a high affinity for the aqueous phase. Using these assumptions it is possible to calculate the bound concentration as a function of the free concentration using the following equation. The derivation of the equation has been presented previously.[7]

$$K_c = \frac{K}{K'} - 1 \times 10^6 / DOC$$

K - binding constant ([g compound/gDOC]/[g compound/g water])

K^c - sorption constant in distilled water ([g compound/g OV-1]/ [g compound/g water])

K' - sorption constant in presence of humic material

DOC - dissolved organic carbon (mg/L)

A number of techniques have been used by other authors, and these will be summarized briefly. Wershaw et al[4] and Matsuda and Schnitzer[20] have both used the solubility technique. The techniques and interpretation are essentially identical to those used in this research. Means and Wiyajaratne[8] have used an ultrafiltration technique. In this technique a solution of a pesticide in either distilled water or a natural water previously concentrated by ultrafiltration is passed through an ultrafilter or a reverse osmosis membrane. The free pesticide can pass through the membrane, so its concentration can be measured. The difference between the concentration in the ultra-filtered water and the concentration inside the ultrafiltration cell is therefore a measure of the bound concentration. Griffin and Chian[11], Hassett[21], and Diachenko[9] have used volatilization measurements to determine the extent of binding of pesticides and pollutants to dissolved humic materials. In these experiments either the rate of gas stripping of a compound or its equilibrium vapor pressure is measured in the presence and absence of humic materials. The results obtained can be manipulated in such a way to determine the percentage of the pollutant bound.

Comparisons Between Methods

We have made some preliminary comparisons of the methods used in this research.[7] In general the dialysis experiments give the most reliable results. These experiments can not, however, be run with all compounds. For example, diethyl-hexyl-phthalate DEHP (MW = 392) would not diffuse through dialysis membranes with molecular weight cut off values of either 1000 or 2000 Daltons. Dialysis tubing with higher molecular weight cut off values than 2000 was unable to effectively retain the humic materials and therefore could not be used. For these reasons it was not possible to run the dialysis experiments on DEHP. In order to compare the results obtained with DEHP with the results for other compounds it was necessary to run some experiments to compare the different methods used. The data is presented in Table I.

The comparisons in Table I using anthracene as the model pollutant showed that the dialysis and sorption techniques compare well. For both Boonton Humic Acid and Pakim Pond Humic Acid the results were not significantly different. The solubility results were significantly low in the case of Boonton Humic Acid (BSHA) and were high in the case Pakim Pond Humic Acid (PPHA). However, all of the measurements were within a factor of three for the different methods, and all showed the BSHA bound more anthracene than PPHA.

The comparisons using DDT as the model pollutant showed a larger discrepancy between the results for the dialysis and sorption techniques than the comparisons for anthracene. For both samples of humic acid the sorption technique gave results that were higher by about a factor of two. The cause of the discrepancy is not clear. Although the methods produce different results, the results are not so different that they prohibit any conclusions. For example, both techniques show that DDT is quite strongly bound by the humic acids, and that the binding is much more extensive than for anthracene. Both techniques also show that Boonton Humic Acid binds DDT more strongly than Pakim Pond Humic Acid.

Table I also shows the comparisons between the solubility and sorption methods for DEHP. Dialysis results were not compared because, as previously mentioned, DEHP would not diffuse through the dialysis membranes. The results from both methods were highly inconsistent and they showed serious discrepancies. Two series of solubility experiments using Pakim Pond Humic Acid gave results which compared quite well. Two series of solubility experiments with Boonton Humic Acid, however, gave results which differed by an order of magnitude. The higher results agreed more closely with the solubility results. The sorption results for both humic acids gave quite high results, which for Pakim Pond Humic Acid were higher than the solubility results by a factor of three. The high values are similar to those found by Matsuda and Schnitzer[20] who used a solubility technique to measure the extent of binding of various phthalates to soil fulvic

TABLE I

METHOD COMPARISONS

ANTHRACENE

	BOONTON HUMIC ACID	PAKIM POND HUMIC ACID
Dialysis	83400 (29300)	12900 (5500)
Solubility	64800 (3200)	41200 (1000)
Sorption	94800 (7300)	12100 (3300)

DDT

	BOONTON HUMIC ACID	PAKIM POND HUMIC ACID
Dialysis	417,000 (66200)	122,000 (14800)
Sorption	751,000 (47600)	199,000 (38800)

DEHP

	BOONTON HUMIC ACID	PAKIM POND HUMIC ACID
Solubility	149,000 (66200)	147,000 (97600)
Solubility*	1,375000 (873,000)	138,000**
Sorption	1,680,000 (1,105,000)	531,000 (327,000)

* Different HA concentrations.
**Only two determinations.
  Values in parentheses are 95% confidence limits.

acid. The binding constant calculated from their data for
DEHP is about 3,000,000. Despite all of these
discrepancies, the results again indicate that more of the
pollutant is bound to Boonton Humic Acid than to Pakim Pond
Humic Acid. Both techniques also show that more DEHP is
bound to the humic acids than anthracene.

Quantitative Binding Measurements

Using the above techniques we have measured binding
constants for a number of compounds to dissolved humic
materials. Some of the results will be summarized in the
following paragraphs.

Figures 1 and 2 show the effect of temperature on the
extent of binding of DDT to both Pakim Pond Humic Acid and
Boonton Humic Acid. The Y axis in these figures is the
amount of DDT bound to the humic acid in nanograms of DDT
per gram of humic acid. The X axis is the free, truly
dissolved DDT in nanograms per liter. This is similar to
the presentation of an adsorption isotherm. If the slope of
the lines is multiplied by 1000 it becomes analogous to a
weight-weight partition coefficient ([g DDT/g DOC]/[g DDT/g
water]). This will be referred to as the binding constant.
A number of things should be noted in these figures. First,
for both Pakim Pond Humic Acid and Boonton Humic Acid a
decrease in temperature results in a significant increase in
the binding constant. The binding constant for Pakim Pond
humic acid increased from 120,000 to 220,000. Boonton humic
acid increased from 410,000 to 700,000. Second, all of the
curves appear to be linear, and all of them pass very close
to the origin. This is similar to the results of numerous
studies of the sorption of organic compounds to sediments in
that a linear isotherm is an adequate model for the data
over most of the relevant concentration range. Third, both
sets of data are quite consistent. This reflects the
relative ease and precision of the dialysis technique.

Figure 3 presents some measurements of the extent of
binding of DDT to dissolved organic carbon in secondary
sewage effluent. The data is somewhat noisy, partially due
to the fact that a portion of the DOC was able to pass
through the dialysis bag. Nonetheless, the experiments
showed that the DDT was bound to the sewage effluent DOC.
The value of the binding constant from this experiment is
76300. The data can be adjusted for the leakage through the
dialysis bag assuming that the DOC which leaked was also
capable of binding DDT. Using this set of assumptions, the
binding constant is 193,700. These two values, at 20 mg/l
DOC, correspond to 60% and 80% bound. Figure 4 is a graph

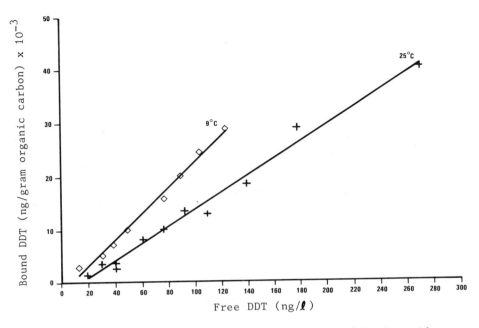

Figure 1.   Effect of Temperature, Pakim pond humic acid.

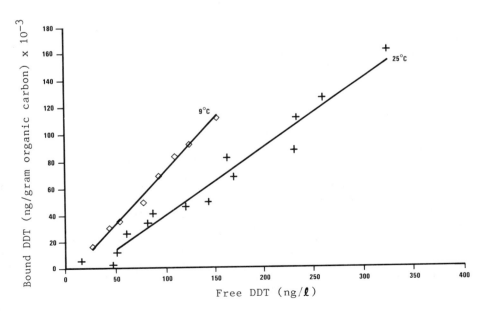

Figure 2.   Effect of temperature, Boonton humic acid.

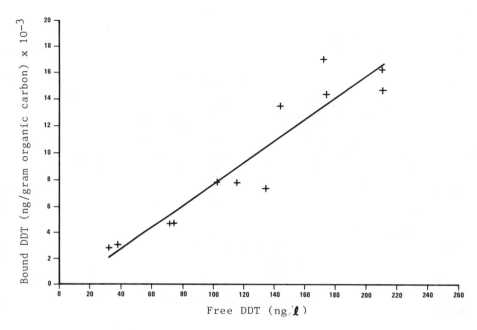

Figure 3. DDT binding by sewage effluent DOC.

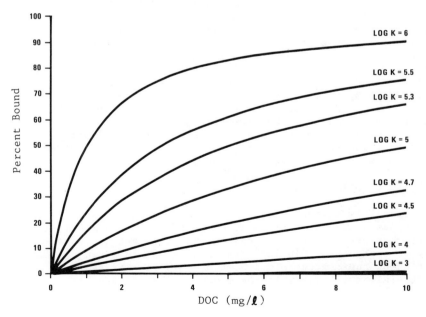

Figure 4. Percent bound vs. DOC for different values of K.

of % bound vs. DOC at various values of the binding
constant. This can be used to convert the constants back to
environmentally useful terms.

Binding of Various Compounds By Pakim Pond Humic Acid

We attempted to determine what properties of a compound
resulted in its being strongly bound to humic materials.
The binding constants of a number of compounds were measured
using dialysis, solubility and sorption techniques. The
solubility technique was used for compounds which were not
radiolabeled. All data was collected at pH = 8.3. The
binding constants were then compared to the octanol/water
partition coefficients for the compounds and the molar
solubilities of the compounds. The data is presented in
Table II. The Kow values were taken from the
literature.[18,22-24] The solubility values were determined
in this research with the exception of DDT and Lindane,
which were taken from the literature. A plot of log Kc vs.
log Kow is presented in Figure 5. The slope of this line is
0.71, the intercept is 0.75 and the value of the correlation
coefficient is 0.9258. The regression is highly significant
(p 0.001).
      It should be noted, that there is considerable latitude
in the data. If the solubility data is used for DEHP and
anthracene instead of the sorption and dialysis data the
slope decreases, the intercept increases and the correlation
is poorer. There is also a large amount of scatter in the
data for the compounds with the lowest association
constants, Lindane and di-n-butyl phthalate. Another source
of error in this regression is in the literature data for
the octanol/water partition coefficient. Values reported by
different authors sometimes vary by orders of magnitude.[20] Despite these shortcomings in the data it is
clear that binding constant is strongly related to the Kow
value for a compound. This data is good news for modelers
in that some of the concepts used in modeling sediment
pollutant interactions may also be applicable to dissolved
humic material - pollutant interactions.

Binding of DDT By Various Humic Materials

Unfortunately there is also some bad news for modelers.
Different humic materials bind compounds to dramatically
different extents, and the reasons for this are unclear.
Figure 6 shows the binding constants of DDT to seven
different humic materials. Some of this data is from a
factorial experiment which has been published elsewhere.[7]
Inspection of this data shows that the humic acids and the

TABLE II

LOG K_C vs. LOG K_{OW} DATA

COMPOUND	LOG K_C	LOG K_{OW}	TECHNIQUE
Fluorene	3.95	3.99	Solubility
Fluoranthene	4.59	5.22	Solubility
DBP	3.70	4.91	Solubility
DCHP	4.72	5.71	Solubility
TCB	4.06	4.27	Solubility
DEHP	5.15	6.69	Sorption
Anthracene	4.60	4.63	Dialysis
Lindane	3.04	3.70	Dialysis
DDT	5.09	6.36	Dialysis

DBP, di-n-butylphthalate
DCHP, Dicyclohexylphthalate
TCB, 1,2,4, trichlorobenzene
DEHP, di-(2-ethylhexyl) phthalate

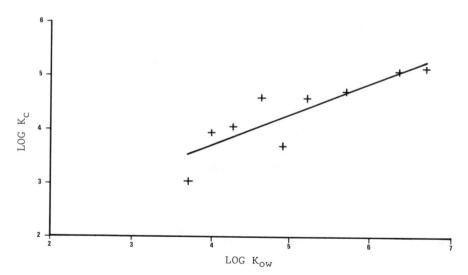

Figure 5. K_{OW} vs. K_C

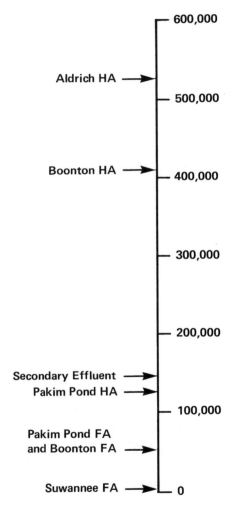

Figure 6. K_C for DDT.

sewage effluent DOC tended to bind DDT most strongly. Two
of the fulvic acids (soluble at pH=1) also bound the DDT,
but to a lesser extent than the humic acids. One fulvic
acid (Suwannee fulvic acid) showed no tendency whatsoever to
bind DDT. This puzzling observation was also true for
anthracene and DEHP. Suwannee fulvic acid is also the only
material used in this research which was subjected to
extensive cleanup procedures. It is possible that the lack
of any binding ability is due to the cleanup procedures, and
not to inherent differences in the fulvic acids from
different sources. A variety of attempts were made to
determine what characteristics of the humic materials
resulted in the differences in binding. The measurements
included % carbon, % ash, % iron, various spectroscopic
techniques, molecular size estimates and pyrolysis gas
chromatography. None of these efforts were successful.
With the data currently available, it is not possible to
predict how strongly a particular sample of humic material
will bind a pollutant without actually measuring the binding
constant.

Conclusions

There is a body of data in the literature which indicates
that dissolved humic materials may play a significant and
previously overlooked role in the behavior of organic water
pollutants. It has been shown that dissolved humic
materials can affect degradation rates and phase transfer
rates for a number of compounds. A number of methods have
been developed in this research and by other researchers
which can make quantitative measurements of the extent of
binding between organic water pollutants and dissolved humic
materials. Hopefully these methods will be used by other
researchers to gain more insight into this phenomenon.
    The data presented here indicates that the extent of
binding for a particular compound is related to the
octanol/water partition coefficient for that compound. This
is very similar to the sorption of compounds from water to
sediment. Compounds with log Kow values less than four
(such as Lindane) will probably not be bound to an
appreciable extent in the environment. Compounds with very
high log Kow values (DDT and DEHP) may be bound to a
significant extent. The extent of binding will depend on
both the concentration of humic material and on the nature
of the humic material. The humic materials used in this
research showed dramatically different affinities for DDT.
The reasons for this are poorly understood and deserve
further study.

Literature Cited

1. Baughman, G.L.; Lassiter, R. R. "Estimating the Hazard of Chemical Substances to Aquatic Life." Cairns, J., Dickson, K.L. and Maki, A.w., eds. American Society for Testing and Materials, Philadelphia, 1978.
2. Mackay, D. Env. Sci. and Tech. 1979, 13, 1218-1223.
3. Neely, W.B. "Chemicals in the Environment" Marcel Dekker, New York, 1980.
4. Wershaw, R.L.; Burcar, P.J; Goldberg, M.C. Env. Sci. and Tech. 1969, 3, 271-273.
5. Porrier, M.A.; Bordelon, B.R.; Laseter, J.L. Env. Sci. and Tech. 1972, 6, 1033-1035.
6. Carter, C.W.; Suffet, I.H., Env. Sci. and Tech. 1982, in Press.
7. Carter, C.W. Ph.D. Thesis, Drexel University, Philadelphia, Pa. 1982.
8. Means, J.C.; Wijayaratne, R. Science 1982, 215, 968-970.
9. Diachenko, G.W. Ph.D. Thesis, University of Maryland, 1982.
10. Hassett, J.P.; Anderson, M.A. Water Research, 1982.
11. Griffin, R.A.; Chian, E.S.K. "Attenuation of Water Soluble Polychlorinated Biphenyls by Earth Materials" EPA Publication, 600/2-80-027, 1980.
12. Perdue, E.M. Presented at the Symposium on Terrestrial and Aquatic Humic Materials, University of North Carolina, November, 1981.
13. Landrum, P.F.; Geisy, J.P. "Advances in the Identification and Analysis of Organic Pollutants in Water," Keith, L. ed. Ann Arbor Science, Michigan, 1981.
14. Hassett, J.P.; Anderson, M.A., Env. Sci. and Tech. 1979, 13, 1526-1529.
15. Eisenreich, S.J. Presented at the Kansas City Meeting of the American Chemical Society, September, 1982.
16. Zepp, R.G.; Baugham, G.L; Schlotzhauer, P.E. Chemosphere 1981, 10, 109-117.
17. Laversee, G.J. Presented at the Symposium on Terrestrial and Aquatic Humic Materials, Chapel Hill, North Carolina, November, 1981.
18. Yalkowsky, S.H.; Valvani, S.C., J. Pharm. Sci, 1980, 69, 912-922.
19. May, W.E.; Wasik, S.P.; Freeman, D.H., Anal. Chem., 1978, 50, 997-1000.
20. Matsuda, K.; Schnitzer, M. Bull. Env. Tox. Contam. 1973, 6, 200-204.
21. Hassett, J.P., personal communication.
22. Callahan, M.A. et at "Water Environmental Fate of 129 Priority Pollutants" EPA Publication 440/4-79-029a, 1979.

23. Chiou, C.T.; Schmedding, D.W.; Manes, M., Env. Sci. and Tech., 1982, 16, 4-10.
24. Chiou, C.T.; Schmedding, D.W., "Test Protocols for Environmental Fate and Movement of Toxicants" Zweig, G. and Beroza, M. eds. Association of Official Analytical Chemists, Washington, D.C., 1981, Chapter 3.

RECEIVED May 3, 1983.

A Comparative Study of the Relationships Between the Mobility of Alachlor, Butylate, and Metolachlor in Soil and Their Physicochemical Properties

C. J. SPILLNER, V. M. THOMAS, and D. G. TAKAHASHI
Stauffer Chemical Company, Mountain View Research Center,
Pesticide Metabolism Section, Mountain View, CA 94042

H. B. SCHER
Stauffer Chemical Company, DeGuigne Technical Center, Richmond, CA 94804

The order of the mobilities of alachlor, butylate, and metolachlor in columns of various soils was metolachlor > alachlor > butylate. This correlates directly with the water solubilities and inversely to the adsorption coefficients and octanol/water partition coefficients of these compounds. Diffusion of these compounds in soil thin-layers was as follows: butylate > alachlor > metolachlor, which correlates directly with the vapor pressures of these compounds. Significant soil properties affecting diffusion appeared to be bulk density and temperature. Soil moisture is also probably important, but its effect on the diffusion of these compounds was not determined.

The physicochemical properties of a pesticide and its interaction with soil greatly influences both its mobility and biological availability in a soil environment (1). Reviews on this subject have been published by Goring and Hamaker (2) and Greenland and Hayes (3).

The objectives of this study were to (a) determine the mobilities of the herbicides, alachlor (2-chloro-2',6'-diethyl-N-(methoxymethyl)acetanilide), butylate (S-ethyl diisobutylthiocarbamate), and metolachlor (2-chloro-N-(2-ethyl-6-methyl phenyl)-N-(2-methoxy-1-methyl ethyl) acetamide in the laboratory using soil leaching columns and soil thin-layer vapor diffusion techniques, (b) determine their soil adsorption coefficients and other physicochemical properties such as octanol/water partition coefficients, water solubilities, vapor pressures, heats of adsorption and heats

0097–6156/83/0225–0231$06.00/0

of solution and (c) correlate the mobilities and the physicochemical properties of these compounds.

Materials and Methods

Chemicals. Purified, [^{14}C]-labelled alachlor (specific activity
= 17 mCi/mM), butylate (specific activity = 2.54 mCi/mM) and
metolachlor (specific activity = 4.5 mCi/mM) were used in the
leaching, adsorption, and diffusion studies. The radiopurity of
these compounds was greater than 95% as determined by thin-layer
chromatography. All other studies were conducted using analytical grade, non-radioactive material (purity > 95%).

Physical Properties. Octanol/water partition coefficients were
determined following the method described by A. Leo, et al.,
(13). Samples were analyzed by gas chromatography (GC). Water
solubility was determined by equilibration of analytical grade
material with water at constant temperature. Equilibrium was
approached from both under and super saturation conditions and
samples were analyzed by GC. Vapor pressures were determined
by the Knudsen effusion method.

Soils. The physical and chemical properties of the soils used
in these studies are presented in TABLE I. Soils were screened
(500μ) prior to use.

Analytical Methods. Liquid scintillation counting (LSC) was done
using Packard Models 3375 and 3380 Liquid Scintillation Spectro-
meters equipped with automatic external standards. Solid samples
were combusted in a Packard model 306 Sample Oxidizer prior to
LSC analysis.

Soil Thin-Layer Vapor Diffusion. Glass plates (20 x 20 cm) were
covered with soil slurries of Keeton sandy loam and Prairie silty
clay loam to a thickness of 0.75 mm. The plates were allowed to
dry overnight and then 10 mL aliquots of acetone solutions of
[^{14}C]alachlor, [^{14}C]butylate, and [^{14}C]metolachlor (corresponding
to 0.18 μmole and 1.0 x 10^6 dpm) were applied to the soil. The
specific activities of these compounds were all adjusted to 2.54
mCi/mM prior to running these experiments. The treated plates
were evenly sprayed to saturation with distilled water, wrapped
in a plastic film to reduce atmospheric volatility, and then
held in a dark growth chamber maintained at 24°C. At 0-, 12-,
and 24-hour intervals, the plates were removed, and placed under
x-ray film. They were stored in a freezer compartment (-4°C)
for three days before developing the film.

Soil Column Leaching. Glass tubing (diameter = 1 cm) was cut into 50 cm lengths, and one end was plugged with glass wool and
Miracloth®. Air dry soil (percent moisture = 2%, 1%, 4% for Felton, Keeton, Prairie, respectively) was packed into the tubes to
a depth of 30 cm. A small layer of white builders sand was then

TABLE I. Mechanical and Chemical Properties of Soils.[a]

Name	Sand	Silt	Clay	O.M.[b]	pH	SP[c]	BD
Felton sand	90%	6%	4%	3.8%	6.1	28	1.7 g/cc
Keeton sandy loam	66%	26%	8%	4.2%	6.7	38	1.4 g/cc
Prairie silty clay loam[d]	12%	59%	29%	8.2%	6.1	58	1.2 g/cc

a| Analyses performed by Perry Laboratory, Los Gatos, CA., and Soil and Plant Labs, Santa Clara, CA.

b| Organic matter determined by combusting a 2.0 g sample at 550-600°C.

c| Saturation percentage–amount of water to saturate 100 g of soil (approximately twice field capacity).

d| Soil collected from R. Porter Farm, Thurman, Iowa.

added to mark the treatment zone. A weighed 5 cm equivalent of
soil was then treated with 0.5 mL of acetone containing enough
of the ^{14}C herbicides to provide the appropriate treatment rate.
The treated soil was added to the column, and a 15 cm equivalent
of water (12 mL) was added so that the soil was just saturated
(1 mL leachate was collected from the Felton soil columns). After a 3-hour equilibration period, the columns were broken into
2.5 or 5 cm sections. Each section was extracted with 10 mL of
acetone by shaking for 3 hours, the suspension was allowed to
settle, and aliquots of the extracts were analyzed by LSC. All
leaching was done in an environmental chamber held at 26°C. The
average percent recovery of radioactivity was 93.4%.

Soil Adsorption. Soil (2.5 g) and 10 mL of aqueous pesticide solution were combined in 30 mL screw cap (teflon-lined) centrifuge
tubes which were then agitated for 3 hours, in darkness, in a
growth chamber set at 10 ± 1°C, 19 ± 1°C, or 30 ± 2°C. The tubes
were centrifuged and the supernatants were analyzed by LSC. Control experiments included untreated solution/soil mixtures used
for LSC background determrnations and treated solutions without
soil used to determine the extent of pesticide adsorption by the
glass tubes.
    The adsorption solutions from the highest concentrations
runs were extracted with ether. The ether extracts were concentrated and analyzed by TLC. Similarly, the corresponding soils
were extracted with ether and the ether extracts were analyzed
by TLC. Other soil samples were analyzed by combustion in order
to determine directly the amount of adsorbed herbicide.
    The adsorption coefficients (K) were determined using the
equation for the Freundlich adsorption isotherm:

$$C_s = KC_w^{1/n} \qquad\qquad (1)$$

where C_w = equilibrium solution concentration

and C_s = weight absorbed solute/weight solid
                              (at equilibrium).

Least-squares linear regression analysis was performed on the
data.
    The thermodynamic heats of adsorption (ΔH) were calculated
using equation 3, which is derived as follows from the relationship between free energy and the equilibrium constant:

$$\Delta G = -RT\ln(Kd) = \Delta H - T\Delta S \qquad (2)$$

$$\text{therefore, } \ln(Kd) = -\frac{\Delta H}{RT} + \frac{\Delta S}{R} \qquad (3)$$

$$Kd = C_s/C_w \qquad (4)$$

Equation 3 is analogous to the Clausius-Clapeyron equation for
equilibrium of a substance in the vapor and condensed phases (4).

Where ΔG is free energy, R is gas constant (1.987 cal/deg K mole^{-1}), T is degrees Kelvin, and ΔS is entropy. Kd is the distribution constant of the herbicide between the solution phase and the adsorbed phase (equation 4). Thus, least squares linear regression analysis of ln(Kd) vs. 1/T yielded values for heats of adsorption (ΔH) for the herbicides in Keeton soil.

Results

Physical Properties.   Results of these measurements are given in Figure 1.

Soil Thin-Layer Vapor Diffusion.   An example of an autoradiogram obtained from a diffusion experiment is shown in Figure 2. The extent of diffusion of metolachlor, alachlor, and butylate is given in TABLE II. Butylate diffusion increased during the 24

TABLE II.   Diffusion of Alachlor, Butylate, and Metolachlor in Soil Thin-Layers

Compound	Soil	Area of Diffusion (cm^2)				
		0-Dry	0-Moist	6 Hr	12 Hr	24 Hr
Butylate	Keeton	1.1	2.0	8.0	5.3	19
Alachlor		1.1	2.0	2.0	2.3	2.8
Metolachlor		1.0	1.1	1.3	1.3	3.3
Butylate	Prairie	1.3	1.5	11	13	25
Alachlor		1.3	1.3	2.3	2.6	5.7
Metolachlor		1.5	1.5	2.0	1.5	5.2

hour test period while the diffusion of alachlor and metolachlor was rather limited in soil plates which were saturated with water. No diffusion was detected in dry soil. Under all of the conditions considered, the relative degrees of diffusion through a moist thin-layer of soil, was butylate > alachlor ∿ metolachlor. In a second experiment, the extent of diffusion in each soil (saturated with water) was measured as a function of temperature. These results are shown in TABLE III. Temperature had the greatest affect on the diffusion of butylate and less influence on the

TABLE III.   Diffusion of Alachlor, Butylate, and Metolachlor in Soil Thin-Layers at Various Temperatures[a]

Compound	Area of Diffusion (cm^2)					
	Keeton Soil			Prairie Soil		
	13 C	18 C	29 C	13 C	18 C	29 C
Butylate	13	28	36	20	39	20
Alachlor	3.5	4.5	3.8	4.2	7.6	7.6
Metolachlor	3.8	4.5	4.2	3.5	8.6	8.6

[a] Extent of diffusion in 24 hours.

	Metolachlor	Alachlor	Butylate
Structure	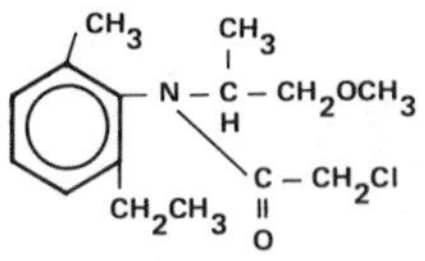	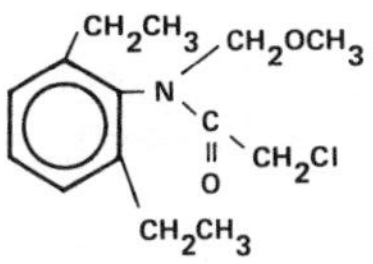	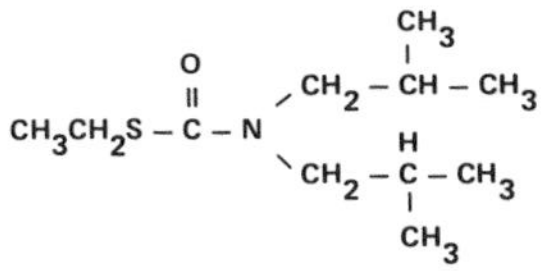
MW	283.8	269.8	217.4
Vapor Pressure (25°C)	3.2×10^{-5} mmHg	3.1×10^{-5} mmHg	1.3×10^{-2} mmHg
Solubility in Water (20°C)	550 ppm	200 ppm	46 ppm
Heat of Solution	Positive*	Positive*	Negative**
Octanol/Water Partition Coefficient (P at 20°C)	1.1×10^3	0.8×10^3	14.0×10^3

* Increase in water solubility with increase in temperature.
** Decrease in water solubility with increase in temperature.

Figure 1. Chemical properties of Alachlor, Butylate, and Metolachlor.

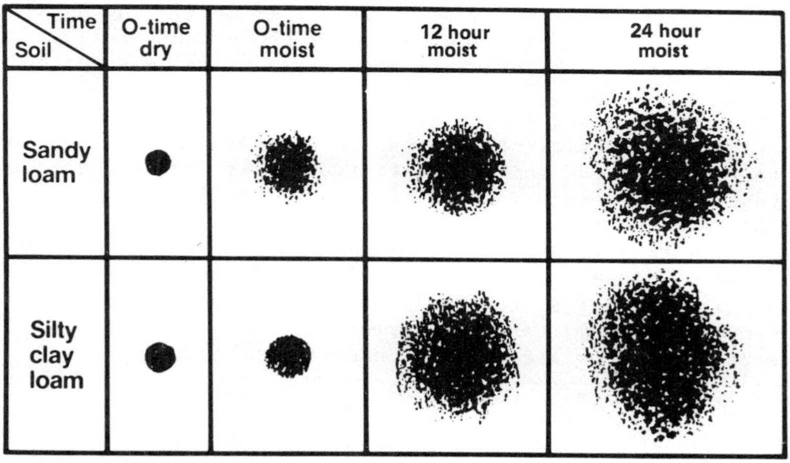

Figure 2. Vapor diffusion of Butylate on sandy loam and silty clay loam.

diffusion of alachlor and metolachlor. In both of these studies
greater diffusion was observed in the Prairie soil in comparison
to the Keeton soil.

Freundlich Soil Adsorption Coefficients. Control experiments indicated that all of the compounds were stable in the stock solutions, in the adsorption solutions, and in the soil during these
studies. A preliminary adsorption run conducted to determine the
time required for equilibration of the herbicides between water
and soil indicated that ca. 3 hours shaking was adequate.

    Results of adsorption experiments for butylate, alachlor,
and metolachlor in Keeton soil at 10, 19, and 30°C were plotted
using the Freundlich equation. A summary of the coefficients obtained from the Freundlich equation for these experiments is presented in TABLE IV. Excellent correlation using the Freundlich
equation over the concentration ranges studied (four orders of
magnitude) is indicated by the r values of 0.99. The n exponent
from the Freundlich equation indicates the extent of linearity of
the adsorption isotherm in the concentration range studied. If
$n = 1$ then adsorption is constant at all concentrations studied
(the adsorption isotherm is linear) and K is equivalent to the
distribution coefficient between the soil and water (Kd), which
is the ratio of the soil concentration (mole/kg) to the solution
concentration (mole/L). A value of $n > 1$ indicates that as the
solution concentration increases the sorption sites become saturated, resulting in a disproportionate amount of chemical being
dissolved. Since n is nearly equal to 1 in these studies, the
adsorption isotherms are nearly linear and the values for Kd
(shown in TABLE IV) correspond closely to K. These Kd values
were used to calculate heats of adsorption (ΔH).

TABLE IV. Adsorption Coefficients for Butylate, Alachlor, and
           Metolachlor in Keeton Soil at Various Temperatures
           Obtained Using the Freundlich Equation.

Compound	T(°C)	K	n	r	Kd[a]
Butylate	10	2.79	1.02	0.99	2.86
	20	3.18	1.01	0.99	3.14
	30	3.28	1.01	0.99	3.26
Alachlor	10	1.65	1.13	0.99	2.00
	20	1.62	1.08	0.99	1.80
	30	1.45	1.09	0.99	1.65
Metolachlor	10	1.98	1.03	0.99	2.10
	20	1.87	1.05	0.99	1.98
	30	1.54	1.06	0.99	1.64

[a] The average distribution constant, calculated from the
     ratio of soil concentration (moles/kg):solution concen-
     tration (moles/L).

<u>Heats of Adsorption</u>. Temperature effects were determined by measuring adsorption at three temperatures. As seen from TABLE IV, the K values vary with temperature such that for butylate, K increases with temperature, while for alachlor and metolachlor, K decreases with temperature. These results indicate that butylate becomes more adsorbed to Keeton soil as the temperature increases while alachlor and metolachlor become less adsorbed as temperature increases. In order to obtain a quantitative measure of these effects, heats of adsorption (ΔH) were calculated as described previously in the Materials and Methods section (equation 3). TABLE IV contains values for the average molar distribution constants (Kd) for butylate, alachlor, and metolachlor which were plotted vs the inverse temperatures ($1/°K$) to obtain the ΔH's shown in Figure 3.

<u>Soil Column Leaching</u>. The distribution of radioactivity from [^{14}C]butylate applied at 4.5 KG/HA and [^{14}C]alachlor and [^{14}C]-metolachlor applied at 2.25 KG/HA and leached with 15 cm of water in Felton sand, is shown in Figure 4. Although all three herbicides are mobile in this soil type, butylate showed less mobility, with 59.6% of the applied raidoactivity found in the upper 10 cm of the column, while 28.4% and 24.3% of the applied ^{14}C was found in the upper 10 cm of the alachlor and metolachlor columns, respectively.

In Keeton sandy loam, mobility was reduced (Figure 4), but 54.6% of the butylate remained in the 5 cm treated area, while 36.3% and 28.2% of the applied alachlor and metolachlor remained in this section.

In Prairie silty clay loam (Figure 4), the mobility of all three herbicides was greatly reduced due to the soils' high organic matter content (8.2%). Most of the applied radioactivity was found in the upper 10 cm of the column for each compound.

Rf values, calculated by dividing the distance moved by the water front by the distance moved by the compounds are given in TABLE V. These values can be used to verify various models. For

TABLE V. Rf of Butylate, Alachlor, and Metolachlor in Various Soil Columns

	Felton	Keeton	Prairie	Calc. Rf[a]
Butylate	0.6	0.33	0.17	0.26
Alachlor	0.9	0.5	0.25	0.41
Metolachlor	0.9	0.8	0.33	0.37

[a] Rf in Keeton soil were calculated as follows (see ref. 17):

$$Rf = (1+(K)(ds)(\frac{1}{p2/3} -1))^{-1} \text{ where}$$

Rf = Rf of the pesticide in the soil column.
K = Freundlich adsorption coefficient from TABLE IV.
ds = bulk density of soil from TABLE I.
p = soil pore fraction (0.476).

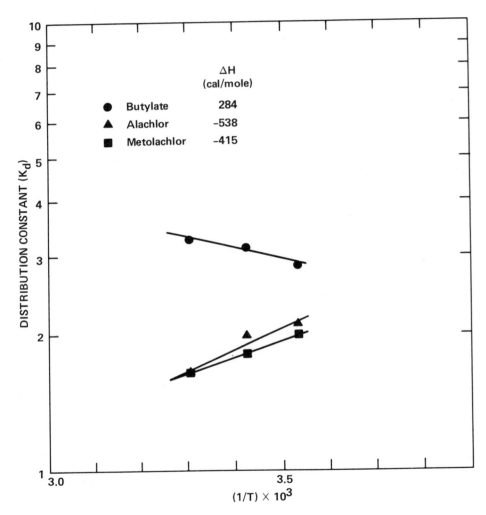

Figure 3.   Heats of adsorption for Butylate, Alachlor, and
Metolachlor in sandy loam soil.

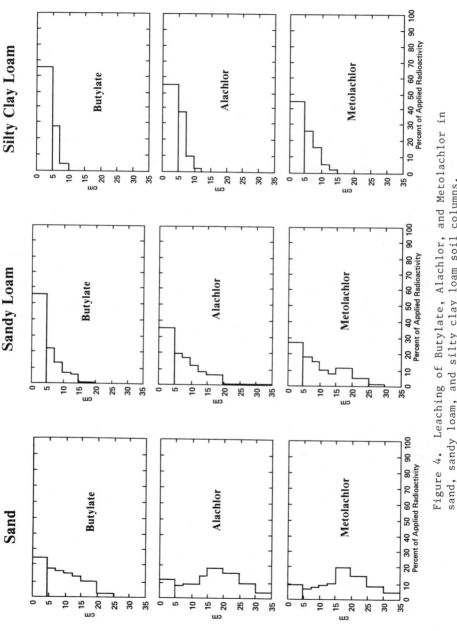

Figure 4. Leaching of Butylate, Alachlor, and Metolachlor in sand, sandy loam, and silty clay loam soil columns.

example, Hamaker (6) gives an equation for converting from adsorption coefficients (K) to Rf values. In view of the relative simplicity of this model, the calculated Rf values presented in TABLE V, (determined from the K values in TABLE V at 20°C) are good approximations of the actual Rf determined in these leaching studies.

Discussion

Soil Diffusion. The results from the soil thin-layer diffusion study of butylate, alachlor, and metolachlor appear to be correlated directly to the vapor pressues and inversely related to water solubilities (TABLE II) in accord with Henry's law of solute/solvent interactions (5). Thus, diffusion is the result of pesticide vapor movement in equilibrium with the liquid phase of the soil environment, rather than diffusion in the liquid phase or movement with the liquid phase. These conclusions are supported by the following observations: 1) movement with the liquid phase can be ruled out in these diffusion studies since in the soil column leaching studies alachlor and metolachlor leached more than butylate, but diffused less; 2) diffusion in the liquid phase is not significant since adsorption to soil organic matter would be expected to play a predominant role, and the results indicate that soil organic matter had no affect on the diffusion of these compounds (i.e., greater diffusion occurred in Prairie soil which had the greatest percent organic matter). The possibility of diffusion occurring in the space between the soil layer and the plastic wrap covering the soil, was ruled out by adding an additional layer of moist soil over the applied herbicides and by observing that the area of diffusion did not change. The various soil properties which appear to be important in the diffusion of these herbicides are soil moisture, soil temperature, and soil bulk density. Although not enough different soils were tested to establish these correlations. The absence of diffusion in air-dry soils was determined in a preliminary experiment and the direct correlation with temperature is clear in TABLE III. There also appears to be an inverse correlation between diffusion and soil bulk density since greater diffusion was observed in the Prairie soil compared to the Keeton soil. The effect of temperature is not surprising in view of the relationship between vapor pressure (p) and temperature (T) shown in equation 5:

$$\log p = A - B/T \qquad (5)$$

where A and B are constants and T is absolute temperature (17). These findings are consistent with the work of Farmer, et al., (18,19) and Igue, et al., (20) who have reported on the importance of these factors (vapor pressure, temperature, soil moisture content, and soil bulk density) and their effect on the diffusion of pesticides in soil. In addition to water solubility,

these are some of the important factors which must be considered
when developing a comprehensive environmental model which includes pesticide diffusion in the soil.

Soil Mobility. The mobility of these compounds in soil leaching
columns can be directly correlated to their respective water solubilities (TABLE II). In all cases, increasing leaching was observed as follows: metolachlor > alachlor > butylate. Furthermore, soil organic matter appears to be the single most important
soil factor affecting the vertical mobility of these compounds.
This is demonstrated by the slight leaching of all three compounds observed in Prairie soil (OM - 8.2%) in which the differences in mobility are minimal.
     Adsorption of these compounds in the soil is a predominant
factor in their mobilities. A thorough understanding of these
processes results in a better understanding of the mobility of
these compounds in soil.
     The K values recorded in TABLE IV are related to adsorption
such that increasing values indicate greater adsorption. In the
present study, butylate exhibited the largest K values in Keeton
soil and is therefore the most strongly adsorbed of the three
compounds studied. These results indicate that butylate would
be the least mobile of these three compounds in that soil type.
This is consistent with the results from the comparative leaching of butylate, alachlor, and metolachlor in three soil types
(including Keeton soil). In all soils, butylate exhibited the
least mobility. Adsorption properties of pesticides have been
shown to most uniformly correlate with the organic matter content
of the soil (6). Obrigawitch et al., concluded that soil organic
matter was the single most important soil property affecting metolachlor adsorption and mobility (7). In a single test at one
concentration, the adsorption of butylate and alachlor was greater in Prairie loam soil (OM = 8.2%) compared to Keeton soil (OM
= 4.2%). The results from the present study indicate that increased adsorption (corresponding to the increase in soil OM) is
responsible for the decreased leaching observed in the soils with
the greatest OM. Other physical properties of these compounds
which are also correlated with their adsorption properties (K)
are water solubility (S_{H_2O}) and octanol/water partition coefficient (Pow). Evidence for these types of correlation abound in
the literature (8). It is not surprising that these physical
properties are correlated since they all reflect the solution
properties of the compound. In systems where soil organic matter
is the principle soil constituent responsible for adsorption, the
correlation between adsorption and octanol/water partition coefficient is reasonable, since octanol as a sorbant simulates the
soil organic matter (9). Recently, a method has been proposed
whereby all of these physical properties (K, Pow, S_{H_2O}) can be
estimated from the reverse phase-HPLC retention time of a compound (10). This is indicative of the similarity in the physical

processes involved in the partitioning of a compound in an RP-
HPLC column (aqueous/organic phase), partitioning in water/soil
and water/octanol systems, and movement in the soil.

Heats of adsorption are shown in Figure 3 for these compounds. These modest ΔH's probably indicate that hydrophobic
bonding is responsible for adsorption (12). This is consistent
with the non-polar nature of these compounds and the important
role of soil organic matter in adsorption. Soil organic matter
is considered to be hydrophobic (11). One usually obtains negative heats of adsorption for pesticides (12) indicating that heat
is evolved during the process (exothermic). However, with butylate, a positive ΔH was observed, which indicates an endothermic
process; thus, heat was absorbed from the system when butylate
was adsorbed by the Keeton soil. The ΔH observed is also in correspondence with the heats of solution measured for these compounds. Alachlor and metolachlor exhibit positive heats of solution (greater solubility at higher temperatures) while butylate
exhibits a negative heat of solution, so that its solubility decreases at elevated temperatures. The effect of temperature on
adsorption is directly linked to the solution properties of these
compounds at various temperatures.

Butylate probably exhibits a negative heat of solution (and
hence a postive heat of adsorption) due to the hydroponic effect
described by Tanford (14). This effect is caused by the disrupt-
ed water molecules rearranging themselves into a lower energy
state at the hydrophobic surface of the butylate molecule. In
addition, there is probably a negative entropy of solution as the
water molecules find themselves in a more ordered state at the
hydrophobic surface of the butylate molecule (15). The butylate
molecule presents a hydrophobic surface from all directions but
metolachlor and alachlor do not (Figure 5).

Conclusions

The good correlation of the results of vapor diffusion and leaching experiments for butylate, alachlor, and metolachlor with
their physical properties has given support to the value of physical property measurements to predict pesticide movement in the
soil.

Transport of the herbicides by vapor diffusion on moist soil
was shown to be directly related to vapor pressure and inversely
related to water solubility. Transport of the herbicides by
leaching was shown to be inversely related to the Freundlich adsorption coefficient which in turn was directly related to the
octanol/water partition coefficient and inversely related to water solubility (16).

Another interesting result was the observed positive heat of
adsorption for butylate (negative heat of solution) and negative
heat of adsorption for alachlor and metolachlor (positive heat of
solution). This result indicates that at low temperatures (near

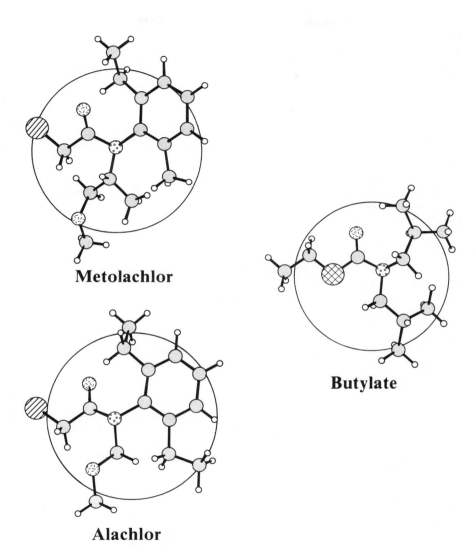

Metolachlor

Butylate

Alachlor

Figure 5.   Computer generated minimum energy configurations of
Metolachlor, Alachlor, and Butylate.

0°C) their relative adsorptivities will converge and that as the
temperature increases, their relative adsorptivities will diverge
as butylate becomes more strongly adsorbed and alachlor and metolachlor become less strongly adsorbed. This result should trans-
late into a reduction of leaching of butylate (compared to alachlor and metolachlor) as the temperature of the soil system is
raised. Thus, the effect of temperature can be handled by an
environmental model for soil mobility by including the heat of
adsorption of the pesticide.

Acknowledgments

The authors would like to acknowledge the contributions of the
following researchers: J.R. DeBaun and L.S. Mullen-Rokita, for
helpful discussions; E.B. Cramer, for assisting with the adsorption measurements; L.-S. Yu-Farina, for the water solubility and
partition coefficient measurements; H. Myers, for the vapor pressure measurements; and R.R. Winter, for running the MACCS molecular structure analyses.

Literature Cited

1. Haque, R. and V.H.E. Freed, "Environmental Dynamics of Pesti-
    cides", Plenum Press, New York (1975).
2. Goring, C.A.I. and J.W. Hamaker. "Organic Chemicals in the
    Soil Environment". Volume 1. Marcel Dekker, Inc., New York
    (1972).
3. Greenland, D.J. and M.H.B. Hayes. "The Chemistry of Soil
    Processes". John Wiley and Sons, New York (1981).
4. Klutz, I., "Introduction to Chemical Thermodynamics", W.A. Ben-
    jamin, Inc., New York (1964).
5. Williamson, .G., "An Introduction to Non-Electrolyte Solu-
    tions", John Wiley and Sons, New York (1967).
6. Hamaker, J.W., in "Environmental Dynamics of Pesticides",
    edited by R.Haque and V.H. Freed, Plenum Press, New York
    (1975) page 115.
7. Obrigawitch, T., F.M. Hons, J.R. Abernathy, and J.R. Gispon,
    Weed Science, 29, 32 (1981).
8. Hassett,J.J., J.C. means, W.L. Banwart, and S.G. Wood, EPA/3-
    90-041 (1980).
9. Briggs, G.G., Proceedings 7th British Insecticide and Fungi-
    cide Conference, pages 83-86 (1973).
10. Swann, R.L., D.A. Laskoski, P.J. McCall,K. Vander Kuy,
    H.J. Dishburger, PEST No. 34, National American Chemical
    Society Meeting, New York, August 23-28 (1981).
11. Ahlrichs, J.L., Chapter 1 in "Organic Chemicals in the Soil
    Environment", edited by Hamaker, J.W. and Goring, C.A.I.,
    Marcel Dekker, Inc., New York (1972).

12. Hamaker J.W. and J.M. Thompson, Chapter 2 in "Organic Chemicals in the Soil Environment", edited by Hamaker, J.W. and Goring, C.A.I., Marcel Dekker, Inc., New York (1972).
13. Leo, A., C. Hausch, and D. Elkins, Chemical Reviews, 71, 525 (1971).
14. Tanford, C., "The Hydrophobic Effect", John Wiley and Sons, New York (1973), pages 20-21.
15. Freed, V.H., R. Hague, J. Vernetti, J. Agric. Food Chem., 15, 1121 (1967).
16. Khan, S.V, "Pesticides in the Soil Environment", El Sevior Scientific Publishing Co., New York (1980), page 41.
17. Spencer, W.F. and M.M. Cliath, "Environmental Dynamics of Pesticides", edited by R. Haque and V.H. Freed, Plenum Press (1975) p. 61.
18. Farmer, W.J., K. Igue, and W.F. Spencer, J. Environ. Qual., 2, 107 (1973).
19. Farmer, W.J., K. Igue, W.F. Spencer and J.P. Martin, Soil Sci. Soc. Amer. Proc., 36, 443 (1972).
20. K. Igue, W.J. Farmer, W.F. Spencer and J.P. Martin, Ibid., 36, 447 (1972).

RECEIVED April 29, 1983.

Mathematical Modeling Application to Environmental Risk Assessments

R. C. HONEYCUTT and L. G. BALLANTINE

Ciba-Geigy Corporation, Agricultural Division, Greensboro, NC 27419

This gives an example of fate modeling in which the risks of an insect growth inhibitor, CGA-72662, in aquatic environments were assessed using a combination of the SWRRB and EXAMS mathematical models. Runoff of CGA-72662 from agricultural watersheds was estimated using the SWRRB model. The runoff data were then used to estimate the loading of CGA-72662 into the EXAMS model for aquatic environments. EXAMS was used to estimate the maximum concentrations of CGA-72662 that would occur in various compartments of the defined ponds and lakes. The maximum expected environmental concentrations of CGA-72662 in water were then compared with acute and chronic toxicity data for CGA-72662 in fish and aquatic invertebrates in order to establish a safety factor for CGA-72662 in aquatic environments.

The major objective of this presentation is to illustrate how an environmental risk assessment of a chemical can be made using mathematical models which are available at the present time. CGA-72662, a CIBA-GEIGY insect growth inhibitor, is used as an example to show how a risk assessment can be carried out using the SWRRB runoff model coupled to the EXAMS fate model.

With any environmental risk assessment of a chemical, there are three factors: 1) The environmental fate of a chemical and 2) the exposure to and 3) the toxicity of the chemical to organisms inhabiting the environment in question.

The environmental fate of a chemical is usually a function of many physical and chemical processes which the chemical may encounter from the time it is applied until it dissipates. Such processes include: Photolysis on surfaces, in solution or in air, hydrolysis, biolysis, oxidation, transport by drift, erosion (runoff) and other means of transport and dissipation. Historically, most risk assessments have emphasized the toxicity of a chemical separately without adequate consideration of the amount of exposure to a chemical which an organism might

0097–6156/83/0225–0249$06.00/0

encounter. However, when one considers first the use pattern
and environmental fate of a chemical and uses these to predict
the amount of exposure of an organism to the chemical, then a
more realistic risk assessment is achieved. For example, a
chemical may be very toxic to fish. However, if the chemical is
degraded rapidly in the environment or adsorbs readily to soil
sediment, it may not pose a significant risk to fish living in
areas adjacent to its application. With the advent of environmental models, one can assess the fate of a chemical and couple
these data to exposure and toxicity data to determine safety
margins for biota in the environment.

CIBA GEIGY Corporation is presently using models as an aid
to data interpretation for risk assessment. Our general philosophy is to use the model as an aid to risk assessment and not
as a predictive tool to eliminate definitive studies. Hopefully, environmental fate models will be useful as a predictive
tool as they become validated.

CGA-72662 will be used as an example to briefly illustrate
an approach to the use of models in environmental risk assessment. CGA-72662 is an insecticide which is being developed for
use on celery in Florida. The celery is usually grown on a high
organic matter muck soil. The recommended application rate is
0.125 lbs. ai/A. A maximum of twelve applications at seven day
intervals may be used for one crop of celery for a total of 1.5
lbs. ai/A. At first glance one might suspect that an environmental hazard might exist from runoff into lakes or ponds adjacent to the application site. As will be demonstrated later,
the SWRRB runoff model was coupled to the EXAMS environmental
fate model to further examine this prospect by predicting the
fate of CGA-72662 and predicting the exposure to aquatic organisms. The results showed very little risk and a high safety
margin for these organisms. Although, the results do not eliminate the necessity to conduct appropriate environmental chemistry studies; the results do give us much confidence that CGA-
72662 used in this manner does not pose a significant environmental risk. The model will also help project the need for
future long-term studies.

Description of the Runoff SWRRB and the EXAMS Models

SWRRB - The Simulator for Water Resources on Rural Basins
(SWRRB) was developed at EPA by R. Carsel and is a modification
of the USDA model CREAMS (1). It was orginally developed to
predict daily runoff volume for small watersheds throughout the
U.S. The basic runoff model is based on the water balance equation:

$$SM_t = SM + P - Q - ET - O - QR$$

SM is the soil moisture at the beginning.
SM_t is the soil moisture t days later.
P is the amount of rainfall.
Q is the amount of runoff.
ET is the amount of evapotranspiration.
O is the amount of percolation below the root zone.
QR is the amount of return flow during the t day period.

Thus, the SWRRB model takes into account many physical processes which contribute to runoff.

The pesticide component of SWRRB takes into account the fate of the chemical applied under field conditions: For example, the amount of pesticide actually reaching the ground after application over a plant canopy is calculated. Further, field dissipation of the chemical by photolysis on leaf surfaces as well as degradation in the soil is accounted for with the pesticide component of SWRRB. Leaching of the pesticide below the top 1cm of soil is also computed and runoff corrected for such losses. Further, adsorption of the pesticide to soil surfaces and sediment is taken into account by SWRRB.

The automated pesticide runoff model consists of a set of FORTRAN programs which calculate the amount of pesticide runoff from input of river basin data, rainfall data, pesticide characteristics, and application data. Table I shows the input requirements for the SWRRB model. Table II shows the output data from the SWRBB model.

Table I
Input data for SWRRB

Pesticide Name
Soil adsorption constant K_d
Washoff fraction
Foliar surface photolysis ($t_{1/2}$ = days)
Soil decay constant (K = days^{-1})
Application efficiency
Initial pesticide on foliage (lbs. ai/A)
Initial pesticide on ground (lbs. ai/A)
Enrichment ratio (pesticide contributed by sediment)
Application day (Julian Calendar)
Application rate (lbs. ai/A)

Table II
Output data for SWRRB

1.   Listing of input data.
     River basin parameters and pesticide characteristics
2.   Average daily temperature and solar radiation
3.   Soil hydraulic properties by layer
4.   Rainfall data
5.   Daily pesticide runoff values
6.   Average monthly and annual values for pesticide runoff

EXAMS - The Exposure Analyses Modeling System was developed
at EPA by Burns, Cline, and Lassiter (2)   The model is based on
the conservation of the mass of a chemical within a dynamic
aquatic environment. The following equation can be used to
mathematically describe the model.

$$- \frac{ds}{dt} = V + P_D + P_S + H + A + M \pm Se + D - L$$

where S = concentration of the chemical in the system
      V = volatilization
      P_D = direct photolysis
      P_S = sensitized photolysis
      H = hydrolysis
      A = breakdown by photo-autotrophs
      M = microbial degradation
      Se = exchanges with sediment reservoirs
      D = dilution
      L = loadings of chemical into system

The particular model can be viewed as composed of several
compartments as shown below for a lake in Figure 1.

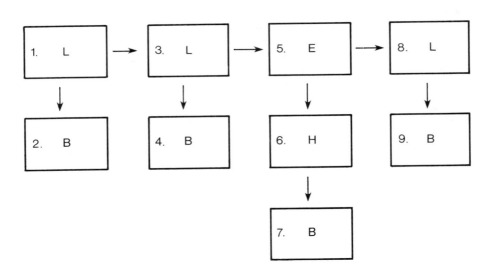

Figure 1. Compartments of EXAMS. Key: L, Littorial, top; B,
Benthic, bottom; E, Ipilimnion, upper layer of water; H,
Hypolimnion, lower layer of water.

Any compartment of the aquatic ecosystem can be represented as a particular volume containing water, particulate matter, biota, dissolved materials, etc. Loadings and exports are represented as mass fluxes across the boundaries of the volume element (processes Se, D and L). Reactive processes are treated as point processes centered within the volume. Thus, the EXAMS model takes into account both physical and chemical processes that affect the environmental fate of a particular chemical.

The automated EXAMS model consists of a set of FORTRAN programs which calculates the fate, exposure and dissipation of the chemical from input environmental data such as: 1) Global parameters (rainfall, irradiance, latitude), 2) Biological parameters (biomass, bacterial counts, chlorophyll), 3) Depths and inlows, 4) Sediment characteristics, 5) Wind, 6) Evaporation, 7) Aeration, 8) Advective and turbulent interconnections, 9) Water flow, 10) Sediment flow, 11) pH and pOH, and 12) Temperature. Also characteristics of the chemical are taken into account such as hydrolysis photolysis, oxidation, biolysis, and volatility.

Table III shows some of the input requirements for EXAMS.

Table III

Input Parameters for EXAMS

1. Compound name
2. Molecular weight (g/mol)
3. Solubility (ppm) mg/L
4. Soil adsorption constant (mg/kg ÷ mg/L)
5. Vapor pressure (torr)
6. Quantum yield
7. Reference latitude
8. Biolysis rate constant (g/hr. cells)
9. Photolysis rate constant (hr.$^{-1}$)
10. Hydrolysis rate constants (e.g., acid hr.$^{-1}$/M)
11. Dissociation constants (acid, base, neutral)

Table IV shows the types of output data from EXAMS:

Table IV

Output Parameters for EXAMS

1. Chemical input data
2. Parameters describing environment
3. Maximum, average and minimum concentrations of chemical at steady state
4. Degree to which each chemical and physical process effects dissipation
5. Distribution of chemical between water, sediment and biota
6. Daily accounting of chemical concentrations in water and sediment
7. Degradation rates for each process
8. Exposure analysis summary including persistence evaluation

Use of SWRRB and EXAMS to Assess the Hazard of CGA-72662 to an
Aquatic Environment

SWRRB

The SWRRB runoff model was used to determine the amount of CGA-
72662 that would runoff of a hypothetical 3.2 acre watershed
with a predominant muck soil type. The following were the SWRRB
input data.

1. Chemical name = CGA-72662
2. Soil adsorption constant = K_d = 49.5
    (The organic matter content of muck soil in Florida is
    about 80%.)
3. The washoff fraction = 1.00.
4. Foliar surface photolysis = $t_{1/2}$ = 1 day.
5. Soil decay constant = 0.008 day^{-1}.
6. Application efficiency = 0.65
    (65% of CGA-72662 reached the ground.)
7. 0 lbs. ai/A on foliage before application.
8. 0 lbs. ai/A on ground before application.
9. 12 applications at 0.125 lbs. ai/A/applications at 7 day
    intervals.
10. River Basin = Watkins 2   3.2 acres.

The runoff for 1974 and 1975 was calculated by SWRRB to be 0.001
lbs. ai/A for each year.
    This runoff figure was then used to calculate the amount of
CGA-72662 that could enter the EXAMS aquatic environments due to
runoff during one season.

SWRBB-EXAMS Interconnections – Calculation for Load Input into
EXAMS Pond

Using the following equations from Reinert (3), the expected
environmental concentration in water (EEC_w) due to runoff into
the EXAMS pond can be calculated:

W_w = Weight of water in EXAMS pond of volume 2 X 10^4 M^3

W_w = 2 X 10^4 M^3 X $\dfrac{1 \text{ ft.}^3}{0.028 \text{ M}^3}$ X 62.4 lb./ft.^{3} = 4.46 X 10^7
lb.

W_s = weight of sediment in pond.

W_s = 6.75 X 10^5 Kg X $\dfrac{1000 \text{ lb.}}{453.6 \text{ kg}}$ = 1.49 X 10^6 lb.

Z_w = load that will partition into water of pond.

$$Z_w = \frac{Z(W_w)}{W_s(K_d)+W_w}$$

Z = load
W_w = weight of water
W_s = weight of sediment
K_d = soil adsorption constant = 49.5

Z = 0.001 lbs. ai/A runoff X 3.2 acres 2 watershed

Z = 0.0032 lbs. load

into $Z_w = \dfrac{0.0032 \ (4.5 \ X \ 10^7)}{(1.5 \ X \ 10^6)(49.5)+4.5 \ X \ 10^7} = 0.0012$ lb. CGA-64250   pond water

is: The Expected Environmental Concentration (EEC_w) in water

$$EEC_w = \frac{Z_w \ X \ 10^6}{W_w} = \frac{0.0012 \ X \ 10^6}{4.46 \ X \ 10^7} = 2.7 \ X \ 10^{-5}$$
ppm

The non point source flow rate (NPSFL) into the EXAMS pond is $5.1 \ X \ 10^3$ kg/hr.

Calculation (3) of the non point source loading rate (NPSLDG) into the EXAMS pond is then:

$$\begin{aligned} NPSLDG &= EEC_w \ X \ NPSFL \ X \ 10^{-6} \ kg/hr. \\ &= 2.7 \ X \ 10^{-5} \ X \ 5.1 \ X \ 10^3 \ X \ 10^{-6} \ kg/hr. \\ NPSLDG &= 1.38 \ X \ 10^{-7} \ kg/hr. \end{aligned}$$

This loading rate is then input into EXAMS pond environment.

The non point source loading rates (NPSLDG) for an Eutrophic Lake or an Oligotrophic Lake can be similarly calculated using the Reinert – (3) Approach.

Use of EXAMS Ponds and Lakes to Determine Environmental Fate of CGA-72662

The following data were input into the EXAMS model to determine the fate of CGA-72662 resulting from runoff (0.001 lbs. ai/A) into ponds or lakes.

1. Molecular weight 166.19 (grams/mole)
2. Solubility 15,000 ppm
3. k_d = 49.5
4. Vapor pressure = 10^{-6} torr
5. Reaction quantum yield = 0.3
6. Direct photolysis rate = $6.93 \ X \ 15^{-2} \ hr.^{-1}$
7. Reference Latitude = 32
8. Hydrolysis (none at pH 5,7,9 – 30-70°C for 28 days)
9. 2nd order rate constant for bottom biolysis = $1.7 \ X \ 10^{-11}$ 100g/hr. cells

10. NPSLDG for pond = 1.38×10^{-7} kg/hr.
11. NPSLDG for Eutrophic Lake = 3.6×10^{-6} kg/hr.
12. NPSLDG for Oligotrophic Lake = 3.6×10^{-6} kg/hr.

The output from EXAMS gives the environmental fate of CGA-72662 and shows what the exposure levels of CGA-72662 are to aquatic organisms inhabiting ponds and lakes adjacent to an application site. These data are shown in Table V.

Table V

Environmental Exposure Levels of CGA-72662

Environment	Maximum Concentration in Water ppm	Maximum Concentration in Sediments ppm	Half-Life in Days	Self-Purification Time Mo.
Pond	1.6×10^{-6}	1.5×10^{-6}	12.6	9
Eutrophic Lake	1.4×10^{-6}	8.5×10^{-7}	61.2	12
Oligotrophic Lake	5.2×10^{-7}	1.7×10^{-7}	4.8	3

The data in Table V indicate that runoff of CGA-72662 from 12 applications would result in extremely low concentrations of CGA-72662 in ponds and lakes. The water column in all cases would contain all of the chemical, the sediment little or no CGA-72662. It follows from these data that exposure of CGA-72662 to aquatic organisms would be low. The data in Table V also shows that CGA-72662 would be persistent only in eutrophic lake environments. After the load is removed, the half-life of CGA-72662 in ponds, eutrophic lakes and oligotrophic lakes was 13, 62, and 5 days respectively. Self purification times were 9, 12, and 3 months respectively.

Table VI shows the final risk assessment of CGA-72662 to aquatic organisms.

Table VI Risk Assessment
CGA-72662

Species	LC_{50} (ppm)	Aquatic Safety Factors (LC_{50}/MEC_w)		
		Pond	Eutrophic Lake	Oligotrophic Lake
Bluegill Sunfish (Lepomis macrochirus)	>90	5.6×10^7	6.4×10^7	1.8×10^8
Rainbow Trout (Salmo gairdneri)	>88	5.5×10^7	6.3×10^7	1.8×10^8
Channel Catfish (Ictalurus punctatus)	>92	5.8×10^7	6.6×10^7	1.8×10^8
Freshwater Invertebrate (Daphnia magna)	93	5.8×10^7	6.6×10^7	9.3×10^8

The toxicity of CGA-72662 to fish and daphnids was determined from aquatic laboratory tests. The LC_{50} was then compared to the maximum environmental concentration of CGA-72662 expected (from EXAMS) in ponds and lakes. The ratio of LC_{50}/MEC_w is called the aquatic safety factor.

Aquatic safety factors ranged from 5.5×10^7 for rainbow trout in ponds to 9.3×10^8 for daphnia in lakes. These data emphasize that exposure levels of CGA-72662 are low and must be taken into account for a risk assessment. Although the persistence of CGA-72662 in eutrophic lakes is relatively long, the exposure is extremely low and of no environmental consequence. Overall, use of SWRRB runoff and EXAMS models show CGA-72662 to be very safe in aquatic habitats when used on vegetables in Florida muck soil.

Limitations of SWRRB and EXAMS Models

No discussion of the use of runoff and environmental fate models would be complete without pointing out their limitations and pitfalls.

SWRRB Limitations and Pitfalls

1. Since the existing watersheds in the model are based on collected field data, choice of application dates are critical since the intensity of rainfall is important. Realistic dates must be chosen to coincide with recommended application times before or during the growing season. E.g., choosing a date just prior to a 4" rain will be a worst case scenario, but may be the wrong time of year.

2. Choice of soil type and adsorption constants are less critical than choice of application dates.

3. The application efficiency must be determined or chosen carefully.

4. The photolysis and soil degradation constants must be realistic for the compound in question. Laboratory or field studies are usually needed to confirm these numbers.

5. Choice of watershed must be realistic and the watershed should have pertinent crops on it.

Table VII shows a sensitivity analysis on the SWRRB model. It can be seen that the intensity of the rainfall is one of the most important parameters affecting runoff.

EXAMS Limitations and Pitfalls

1. Pesticide input data must be accurate and realistic for the chemical in question. E.g., minor changes in input load may result in major changes in output data.

2. The EXAMS model was designated for point source pollution examination. However, modification for non point source pollution can be done.

3. EXAMS may not take into account other important transformation or transport processes that occur in natural aquatic environments. Thus, validation is important.

Table VIII shows a sensitivity analysis on the EXAMS model. Changing the input load dramatically changes the concentration of chemical in both water and sediment. Photolysis rates appear to effect the model less than input loads. Changing the soil type effects the purification time of the system and not so much the water concentrations of the chemical indicating the influence of chemical adsorption to degradation.

Table VII

SENSITIVITY ANALYSIS OF SWRRB MODEL

Soil Type	K_d	Photolysis Half-Life (Days)	Soil Degradation (Days^{-1})	% Onto Soil	Rainfall Intensity	Rainfall in"	CGA-72662 Run-off lbs. ai/A
MUCK	49.5	1.0	0.008	65	INTENSE[a]	14.9	0.023
MUCK	49.5	20.0	0.138	65	LIGHT[c]	7.2	0.001
MUCK	49.5	1.0	0.138	65	LIGHT[c]	7.2	0.001
MUCK	49.5	1.0	0.008	65	LIGHT[c]	7.2	0.001
MUCK	49.5	1.0	0.008	10	LIGHT[c]	7.2	0.001
MUCK	49.5	1.0	0.008	10	INTENSE[a]	14.9	0.004
SAND	0.74	1.0	0.008	65	LIGHT[c]	7.2	0.001
MUCK	14.1	1.0	0.008	65	INTENSE[b]	4.5	0.061
MUCK	14.1	1.0	0.008	65	LIGHT[d]	3.4	0.001

a 1974 - Intense rain period 4.26 inches on day 178 when one application was made (applictions between Julian days 143-220).

b 1975 - Intense rain period (applications between Julian days 240-317).

c 1974 - Light rain period (applications between Julian days 01-12, 300-365).

d 1974 - Light rain period (applications between Julian days 240-317).

Table VIII

SENSITIVITY ANALYSIS OF EXAMS POND MODEL

Soil Type	K_d	Input Load To Pond kg/hr.	Photolysis t 1/2	Max. Conc. in water Of Pond (ppm)	Max. Conc. in Sediment Of Pond (ppm)	Self-Purification Time (mo.)
MUCK	49.5	1.38×10^{-7}*	10 hrs.***	1.6×10^{-6}	1.5×10^{-6}	9 mos.
MUCK	49.5	8.67×10^{-6}**	10 hrs.	1.0×10^{-4}	9.2×10^{-5}	9 mos.
MUCK	49.5	1.38×10^{-7}	100 hrs.	4.2×10^{-6}	3.9×10^{-6}	13 mos.
MUCK	49.5	8.67×10^{-6}	100 hrs.	2.7×10^{-4}	2.5×10^{-4}	13 mos.
SAND	0.74	1.38×10^{-7}	10 hrs.	1.6×10^{-6}	3.4×10^{-7}	1 mo.
SAND	0.74	8.67×10^{-6}	10 hrs.	1.0×10^{-4}	2.2×10^{-5}	1 mo.
SAND	0.74	1.35×10^{-7}	100 hrs.	4.2×10^{-6}	9.2×10^{-7}	3 mos.
SAND	0.74	8.67×10^{-6}	100 hrs.	2.7×10^{-4}	5.8×10^{-5}	3 mos.

* from 0.001 lbs. ai/A runoff.
** from 0.061 lbs. ai/A runoff.
*** t 1/2 = 10 hours rate constant 6.93×10^{-2} hr.$^{-1}$.

Table IX is a summary of the sensitivity of SWRRB and EXAMS to change in inputs. These data are taken from Tables VII and VIII. It can readily be seen that SWRRB is sensitive to rainfall intensity while EXAMS is sensitive to input load changes.

Table IX

EFFECTS ON SWRRB AND EXAMS DUE TO SENSITIVITY

Parameter Changed	Amount of Change	Effects (Fold Change)	
		Runoff lbs. ai/A	Conc. in EXAMS POND (ppm)
Soil Type K_d	67X*	NONE	NONE
Total Rainfall	5X	NONE	–
Intensity of Rainfall	–	23X↑	–
Soil Degradation Rate	17X	NONE	–
Photolysis Rate	20X	NONE	–
Photolysis Rate	10X	–	3X↑
Input Load From Runoff	61X	–	160X↑

$$* \quad \frac{49.5}{0.74} = 67$$

Summary

1. Environmental models which are accessible today can be used for exposure assessment of pesticides.

2. For a realistic risk assessment, the environmental fate, exposure levels and toxicity of the compound must be considered in an integrated fashion.

3. The SWRRB runoff model coupled to the EXAMS fate model can be used to predict exposure levels of chemicals to aquatic organisms. Safety factors can then be calculated.

4. Limitations do exist with each model. Care must be taken to describe both the environments and chemical characteristics in a realistic manner.

Acknowledgments

Assistance of Dr. Bob Carsel at EPA in Athens, Georgia is gratefully acknowledged. Dr. Carsel was instrumental in getting our models to a practical state of usage at CIBA-GEIGY.

Literature Cited

1. Carsel, R. F. Pesticide Runoff Simulator User's Manual, Computer Sciences Corporation, 1980.

2. Lassiter, R. R.; Baughman, G. L.; Burns, L. A., State-of-the-Art in Ecological Modeling, 1978, 7 219-245. Int. Soc. Ecol. Mod., Copenhagen.

3. Reinert, J. L., "Estimating the Maximum Concentration of Pesticides in the Environment as a Consequence of Specific Events" October 1, 1980, Environmental Fate Branch, U.S. EPA.

RECEIVED April 29, 1983.

Application of the Preliminary Pollutant Limit Value (PPLV) Environmental Risk Assessment Approach to Selected Land Uses

DAVID H. ROSENBLATT, MITCHELL J. SMALL, and
ROBERT J. KAINZ

U.S. Army Medical Bioengineering Research & Development Laboratory,
Fort Detrick, Frederick, MD 21701

The site-specific Preliminary Pollutant Limit Value (PPLV) process for environmental risk assessment has been applied to contaminated land areas in order to determine what use might be made of them at various levels of contamination. The process involves examination of the potential for each chemical of concern to proceed from the soil or water, through defined pathways, to man or other target organisms. Each pathway is treated as if it consisted of a series of compartments at equilibrium, except that the exposure of man to the last of these compartments is handled as a consumption rate process. The best available toxicological information is used to estimate an acceptable daily dose, (D_T), for human (or other organism) exposure to each compound. This value is used to calculate levels of the compound in the soil or water such that D_T is not likely to be exceeded during the course of specified categories of human activity. A PPLV is derived from consideration of the D_T along with the probable exposure level. Four specific examples of the use of the PPLV concept are described to illustrate how the concept is applied in real world situations. Soil and water PPLVs are developed for several compounds. These PPLVs vary according to envisioned scenario; for example, subsistence farming, residential housing, hunting, fishing, and industrial or timbering operations. Each scenario entails one or more exposure pathways. The PPLVs so derived allow for various options for cleanup or restriction of land use, such that public health will not be jeopardized by residual contamination. Each potential cleanup effort represents a different level of hazard reduction. The PPLV concept facilitates decisions as to the effective use of limited dollars to clean a site to a level of intended use.

The US Army, which for several years has had responsibilities for renovating contaminated tracts of land, has developed a conceptual framework that can accommodate a variety of decision-making processes and models to respond to the question, "How clean is clean?" This approach focuses on determining acceptable pollutant residue levels as goals for remedial action. It recognizes that potential land use, courses of remedial action, the nature and extent of contamination, and the population at risk are all considerations that may affect those goals. Despite its flexibility, the Army methodology can be described well in terms of referenced scientific estimation methods or correlations and in terms of recently developed paradigms. The Army's Preliminary Pollutant Limit Value (PPLV) concept (1,2,3), a decision tool, is being used and continually improved.

The site-specific PPLV process involves examination of the potential for each chemical of concern to proceed from its point(s) of origin in the soil or water, through defined pathways, to the target organism, typically man (Figure 1). For human targets, the compartments along the pathway are assumed to be at equilibrium, except that human exposure is handled as a rate process (Figure 2). It must be assumed that the compound of interest does not decompose and that if decomposition does occur, the hazard is reduced rather than increased. In cases where this does not hold true and where products pose serious problems, individual detection, identification, and evaluation should be undertaken to address the decomposition products, treating them as new compounds.

The best available toxicological information is used to estimate an acceptable daily dose, D_T, for human exposure to each compound. This is then used to calculate levels of the compound in the soil or water such that D_T is not likely to be exceeded during the course of specified categories of human activity. A PPLV is derived from consideration of the D_T along with the probable exposure level. PPLVs vary according to envisioned scenario, e.g., subsistence farming, residential housing, hunting, fishing, and industrial or timbering operations. Each scenario entails one or more exposure pathways. PPLVs permit assessment of the various options for cleanup or restriction of land use, such that public health will not be jeopardized by residual contamination. Each potential remedial action represents a different level of hazard reduction. The PPLV concept contributes to cost-effective decisions on the use of funds for such remedial actions, in accordance with intended levels of use.

The PPLV process has been applied in several contexts. Each application has revealed new aspects that had not been considered previously (Table I). Nevertheless, the examples share one characteristic common to toxic chemical risk analysis; an acceptable exposure level must be combined with a relationship between source concentration and estimated degree of exposure. This concept has been published previously(1,2,3);

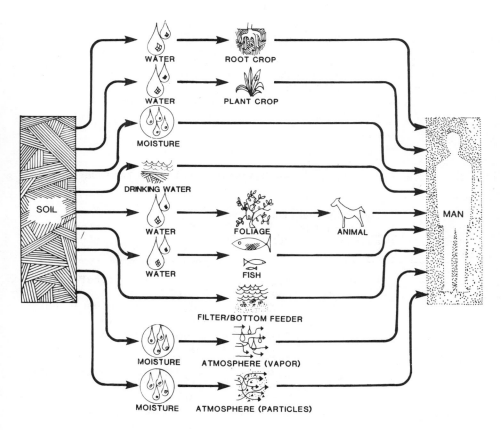

Figure 1.   Pathways from soil via water, plant, and animal compartments to man.

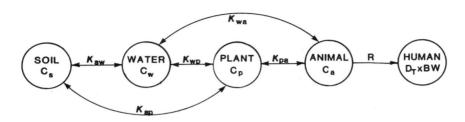

Figure 2. The pathway from soil via water, plants, and animals
to man. In this fate model, the acceptable daily dose of
toxicant, D_T, can be obtained from seven sources of literature
information. The equation for acceptable daily dose is:

$$D_T = \frac{R \times Ca}{BW} = K$$

Table I. PPLV Studies Related to Selected Land Uses

Site (Ref)	Scenarios and Associated Pathways		Compounds of Potential Concern at Site
	Scenarios	Pathways	
Alabama AAP (4)	Subsistence agriculture	Vegetable consumption	2,4,6-Trinitrotoluene (TNT)
		Livestock consumption	2,4-Dinitrotoluene (DNT) N-Methyl-N,2,4,6-tetra- nitroaniline (Tetryl)
		Dairy consumption	1,3,5-Trinitrobenzene (TNB) 1,3-Dinitrobenzene (DNB) Diphenylamine
		Soil ingestion	Aniline N,N-Dimethylaniline Nitrobenzene
	Residential housing	Vegetable consumption	Nitrocellulose[a] Lead[b]
		Soil ingestion	
	Apartment housing	Soil ingestion	

Table I--continued

	Industrial	Dust inhalation	
	Hunting	Meat consumption	
	Timber harvesting	Dust inhalation	
Savanna ADA (5)	Recreational fishing	Sediment to bottom feeders	TNT DNT TNB
	Downriver drinking water supply	Dry lagoons leaching to river	Hexahydro-1,3,5-trinitro-1,3,5-triazine (RDX)
Gratiot County Landfill (6)	Subsistence agriculture	Livestock (soil)	Polybromobiphenyls (PBBs)
	Residential housing	Water ingestion	
	Industrial	Dust inhalation or water ingestion	
Bangor Naval Submarine Base	Recreational fishing	Fish and shellfish consumption	TNT RDX Picric acid[c] Picramic acid[c] Propylene glycol dinitrate (PGDN)

a. Not a toxic hazard.
b. Lead presents some complex problems. These were discussed at length by two of the present authors (Ref. 4).
c. Picric and picramic acids are strongly ionized. Not enough is known of their bioconcentration behavior to permit calculation of a bioconcentration factor; hence, they will not be discussed further in the present report.

the present report concentrates on its use in deriving limit
values associated with selected land uses.

Acceptable Daily Doses (D_T)

The acceptable daily dose of a toxicant (in mg/(kg x day)), D_T,
relative to chronic human health effects, is central to PPLV
calculations. Table II lists seven sources of information from
which D_T values may be drawn. From this, it is seen that, if
there is available an ADI (Acceptable Daily Intake) value
originating with the World Health Organization (7), then that
figure should be used as D_T.
    A second excellent, but limited, source of information is
the list in the National Interim Primary Drinking Water Regulations; its MCL (Maximum Concentration Level) values (8) are
directly convertible to D_T values (1,2,3) by applying the factor
weight of water consumed ÷ body weight:

$$D_T = MCL/35$$

    A third generally accepted source of values is the collec-
tion of TLVs (Threshold Limit Values) published by the American
Conference of Governmental Industrial Hygienists (9) and
utilized by the Occupational Safety and Health Administration.
Conversion of these to D_T (10) involves three factors: The
first is division by 7/5 (= 1.4) to convert from a normal 5-day
workweek to a 7-day exposure week. The second is division by
100; this allows for exceptionally sensitive individuals, who
would not normally be part of the work force, and takes into
consideration the completely involuntary and unsuspected nature
of the exposure. The third factor converts from TLV (expressed
in mg m^{-3}) to a total dose; the breathing rate, RB', for a 70-kg
person (BW = body weight) doing light work is taken as 12.1
m^3/8-hr day (10). Thus,

$$D_T = (TLV \times RB')/(140 \times BW) = TLV/810$$

TLVs must be used cautiously to preclude the effects of hidden,
inapplicable assumptions.
    A fourth official source of values is the Food and Drug
Administration, whose guidelines find occasional use in deriving
PPLVs. For example, for a given compound, where the guideline
is for concentration in beef fat,

$$D_T = \frac{(Meat\ consumption\ rate)(Fraction\ fat\ in\ meat)(guideline\ value)}{BW}$$

    The remaining listed sources from which D_T might be derived
involve animal experiments. Although similar experiments were
the ultimate source of the first four methods for calculating

Table II. Information Sources from which to Derive Values of Acceptable Daily Doses (D_Ts) of Toxic Pollutants for Human Beings (in Order of Priority)

Input Information	Calculation Required	Reference
Existing Standards		
Acceptable daily intake (ADI)	None	WHO ([7])
Maximum concentration level (MCL) in drinking water	Adjust for water consumption level	EPA ([8])
Threshold limit value (TLV) for occupational exposures	Use factors for breathing rate, exposure time, safety factor of 10^{-2}	ACGIH ([9],[10])
FDA guidelines for concentrations in foods	Use factors for consumption of particular foods	FDA
Experimental Results in Laboratory Animal Studies		
Lifetime no-effect level (NEL_L)	Use safety factor of 10^{-2}	([11])
Ninety-day no-effect level (NEL_{90})	Use safety factor of 10^{-3}	([11])
Acute toxicity (LD_{50})	Use safety factor of 1.155×10^{-5}	([1],[2],[3])

D_T, the toxicological experiments referred to here have not gone through the process of extrapolation, evaluation, and consensus. Thus, they are used only in the absence of better data. The no-effect level (NEL) from a chronic or lifetime study in a laboratory animal is diminished by a factor of 100 (11), i.e.,

$$D_T = NEL_L \times 10^{-2}$$

The widely accepted safety margin of 10^{-2} should be sufficient to allow for interspecies differences and especially susceptible individuals or groups within the population.

The no-effect level from a subchronic (90-day) study is assigned an additional safety factor of 10 because of the shorter period of exposure (11). Hence

$$D_T = NEL_{90} \times 10^{-3}$$

The most likely toxicity value to be found in the literature is the LD_{50} (dose lethal to 50% of the animals) for some laboratory species, usually rat or mouse. This value may be obtained by plotting on probit paper the fraction of experimental animals killed against the acute dosage. There is seldom enough information to permit extrapolation to a dosage at which only a very small (e.g., 1%) fraction of the animals would be killed, much less to an acceptable risk level. Handy and Schindler (12), however, assume a safe limit for the maximum body concentration of a toxic substance to be $5 \times 10^{-4} \times LD_{50}$. Based on experimental studies, they also assume a biological half-life of 30 days, which implies a disappearance rate of 2.31% per day. If the daily intake of the toxic substance is made equal to the daily disappearance rate at the safe concentration limit, then that safe concentration is maintained. Thus,

$$D_T = 2.31 \times 10^{-2} \times 5 \times 10^{-4} \times LD_{50} = 1.155 \times 10^{-5} \times LD_{50}$$

This is the least desirable method of estimating D_T but may be the only available method when new or relatively unfamiliar compounds are being dealt with.

Carcinogens pose a special challenge; although characterization of certain compounds as suspected carcinogens might be agreed upon by most researchers, there is no consensus among scientists regarding a suitable mathematical model for carcinogenesis; neither is there an accepted "safe" level for a carcinogen. For the time being, the authors regard a risk level of one cancer death per hundred thousand lifetime exposures to be an acceptable criterion for carcinogenic pollutants that do not threaten a large population. Such criteria have been published by the U.S. Environmental Protection Agency for drinking water pollutants (13).

Partition Coefficients

Assume that between any two adjacent media (such as soil and water or water and crops) the pollutant is partitioned in a perfectly constant manner, e.g.,

C_s = Acceptable concentration in soil
$C_w = K_{sw}C_s$ = Acceptable concentration in water
$C_p = K_{wp}C_w$ = Acceptable concentration in plant(s), dry weight basis
$C_a = K_{pa}C_p$ = Acceptable concentration in meat animal

Also,

$$C_p = K_{sp}C_s = K_{sw}K_{wp}C_s$$

The K values are partition coefficients. The assumption that these are real constants is seldom completely true, of course, because equilibrium is rarely achieved and because the equilibrium ratios generally are not the same for all concentration levels. Moreover, it is difficult to find the needed information, and one must often accept a single literature value as typical of a given intermedia transfer. When the organic content of the soil is known or can be accurately estimated, one can usually derive K_{sw} from a compound's aqueous solubility, S, or its octanol/water partition coefficient, K_{ow} (14). Values of K_{pa}, namely "bioconcentration factors" between feed and meat animals (15,16), can also be derived from S or K_{ow}. Bioconcentration factors between water and fish are well documented (14). A considerable weakness exists in our perception of the proper estimates to use for partition coefficients between soil and edible crop materials. Thus, at one time, two of the present authors used a default value of K_{sp} = 1 for munitions compounds that are neither very soluble in water nor very insoluble (4); at another time, a value of K_{wp} was assumed for compounds with very low values of K_{sw}, i.e., polybromobiphenyls (6).

Scenarios, Single Pathway PPLVs (SPPPLVs), and PPLVs

For each category of land or water body use, one may envision a simplified scenario. In each scenario, only those activities most likely to lead to toxic exposures are considered. For example, in the industrial scenario, indoor workers would not be exposed to levels of dust bearing high concentrations of soil contaminants; outdoor workers who stir up dry soil with heavy machinery, however, could expect to inhale contaminant-laden dust. A scenario could involve more than one exposure pathway. Thus, the industrial worker might drink water from a contaminated well, in addition to breathing contaminated dust; these exposures might represent not only different pathways but different sources.

An SPPPLV usually involves D_T, one or more K values, and a rate of ingestion, inhalation, or other contaminant transfer factor. Additional factors may be included to account for effects that modify intensity or time of exposure. Examples of general equations of this sort have been published previously (1,2,3); those useful in the present examples have been modified as required.

Where pathways have their origin in the same medium or have a common point of intersection, a simple calculation is used to adjust the concentration at the origin or the intersection such that SPPPLVs taken together provide the target organism, usually humans, with an exposure providing exactly D_T. Thus, for SPPPLVs via three pathways from the same source:

$$PPLV = [1/(SPPPLV)_1 + 1/(SPPPLV)_2 + 1/(SPPPLV)_3]^{-1}$$

If several independent sources of the pollutant have to be considered, the individually calculated PPLVs must be reduced to some arbitrary combination consistent with the D_T.

Physicochemical Properties

For the studies summarized in Table I and discussed in the following sections of the text, physicochemical properties (including partition coefficients) were collected from a variety of reference documents or estimated according to available equations (Table III). Acceptable daily doses were calculated from toxicological data (Table III). When more than one equation was available, judgment was used to determine which to apply. Table III excludes those contaminants footnoted in Table I. A default value of 1.0 was adopted for K_{sp} for the first nine compounds of Table III (4). For PBBs, the value of log K_{oc} was calculated from the solubility in creek water (7.96 x 10^{-4} μM), according to the equation (1,3):

$$\log K_{oc} = -0.557 \log S + 4.277 = 6.003$$

With an assumed 2% organic carbon content in soil, K_{sw} = 5 x 10^{-5} for PBBs; if K_{wp} is assumed to be 5, $K_{sp} = K_{sw}$ x K_{wp} = 2.5 x 10^{-4}.

Alabama Army Ammunition Plant (4)

The 5,168-acre Alabama Army Ammunition Plant tract, on the banks of the Coosa River in Talladega County, AL, is 4 miles north of Childersburg, AL (4). Plant operations, between 1942 and 1945, left residues from the manufacture of diphenylamine, TNT, DNT, and tetryl. Some of these compounds have been found on the site, and others are suspected. The shallow water table, draining to the Coosa River, is probably contaminated, but only deeper, uncontaminated aquifers would be used as the source for

Table III. Physicochemical Constants and Acceptable Daily Doses (D_T) of Soil and Water Contaminants[a]

Contaminant[b]	S^c (mg/L)	Log K_{ow}^d	k_2^e (day^{-1})	K_{pa}^f	BCF^g	D_T^h (mg/kg/ day)
TNT	124[i]	1.84[j]	5.1[k]	6.7×10^{-3}	14.7[l]	1.4×10^{-3}
DNT	273[i]	1.98[m]	4.6[k]	4.3×10^{-3}	--	3.2×10^{-5}
Tetryl	35[n]	--	--	1.4×20^{-2}	--	1.8×10^{-3}
TNB	32[i]	1.18[m]	9.6[k]	1.4×10^{-2}	--	5.8×10^{-3}
DNB	370[i]	--	--	3.6×10^{-3}	--	1.2×10^{-3}
Diphenyl- amine	36[o]	--	--	1.3×10^{-2}	--	2.0×10^{-2}
Aniline	35,000[o]	--	--	2.8×10^{-4}	--	1.2×10^{-2}
N,N-Dimethyl- aniline	16,000[o]	--	--	4.4×10^{-4}	--	3.0×10^{-2}
Nitrobenzene	1,780[p]	--	--	9×10^{-3}	--	6.2×10^{-3}
RDX	44[i]	0.87[q]	12.9[k]	--	4.2[r]	1.0×10^{-3s}
PBBs	5×10^{-4t}	--	--	--	--	3.73×10^{-4u}
PGDN	1,300[v]	0.90[w]	--	--	2.8[l]	2.5×10^{-4x}

a. Only data useful for the present report are given here.
b. See Table 1 for abbreviations.
c. Solubility.
d. Log octanol/water partition coefficient.
e. Depuration rate constant for fish (17).
f. Basic data for equations presumably involve contaminant con-centrations in dry feed. Experiments were carried out long enough for a steady concentration in fat to be reached for any concentration in feed. K_{pa} = BF x 0.3 for cattle. For all calculated values of BF shown in this table, the equation used is log BF = 1.2-0.56 log S, (when S must be expressed in mg/L) (15).

Table III continued on next page

Table III--continued

g. BCF = Concentration in whole fish/concentration in water.
h. D_T = acceptable daily dose, as discussed in text. Sources of D_T are provided for the first 9 compounds in Reference (4); the value for DNT is based on a criterion of one excess cancer death in 10^5 lifetime exposure.
i. Reference (18).
j. Calculated from value for TNB (19) by method of Reference (14).
k. Log k_2 = 1.47-0.414 log K_{ow}; k_2 in day^{-1} (17).
l. Log BCF = 0.76 log K_{ow} -0.3, from Reference (14).
m. Reference (19).
n. Reference (20).
o. Reference (21).
p. Reference (15).
q. Reference (22).
r. Average value for 27-day studies (23).
s. Reference (24).
t. For creek water; solubility in distilled water is 9 times lower (25).
u. Based on FDA guideline of 0.3 mg/kg in beef fat (26), 30% fat in beef, and consumption of 0.29 kg/day,

$$D_T = \frac{0.29 \times 0.3 \times 0.3}{BW} = 3.73 \times 10^{-4} \text{ mg/kg}$$

v. Reference (27).
w. Estimated according to Reference (14).

x. Estimated from TLV (9) of 0.2 mg/m^3 according to equation in text.

$$D_T = \frac{(\text{Breathing rate for 8-hr work day}) \times (\text{TLV}) \times (5 \text{ day work week})}{(\text{Body weight}) \times (7 \text{ day work week}) \times (\text{safety factor})}$$

$$= \frac{12.1 \text{ m}^3 \times 5 \text{ days} \times 0.2 \text{ mg/m}^3}{70 \text{ kg} \times 7 \text{ days} \times 100}$$

water supplies. Hence, groundwater contamination was not considered in PPLV scenarios.

Consumption of livestock and dairy products, as well as ingestion of soil by children, were considered for the subsistence agriculture case, but the SPPPLVs were significantly higher than for vegetable consumption; similarly, soil ingestion by children was considered for the residential housing scenario.

For the subsistence farming and residential housing scenarios, the only significant pathway for the nine organic compounds of concern would be that from soil via vegetable consumption. The SPPPLV (and also, in this case, the PPLV) would be:

$$C_s = \frac{BW \times D_T}{\text{Vegetable Consumption (dry weight)} \times K_{sp}} =$$

$$\frac{70 \ D_T}{0.0734 \text{ kg} \times 1} = 953 \ D_T$$

On the other hand, if the land were to be used for apartment housing, the growing of significant amounts of vegetables would not be expected. Here, the only significant pathway was adjudged to be via ingestion of soil by children, with an estimated consumption of 0.1 g (10^{-4} kg), per 12-kg child per day. The pathway-related equation for this situation is:

$$C_s = \frac{12 \ D_T}{10^{-4}} = 1.2 \times 10^5 \ D_T$$

For the industrial setting, the outdoor worker scenario, as mentioned above, represents a worst case. Owing to wind and weather conditions, one assumes that the worker is exposed to dust only 50% of his approximately 225 workdays. The maximum dust concentration is the normal nuisance dust TLV (9) of 10^{-5} kg/m^3, breathed by a 70-kg adult at a rate of 12.1 m^3 per 8-hour workday (4). A factor of 10 is introduced (see equation below) to account for a more robust worker population than the general population. From these assumptions, the calculation is

$$C_s = \frac{365 \times 70 \times 10 \times D_T}{10^{-5} \times 225 \times 0.4 \times 12.1} = 2.35 \times 10^7 \ D_T$$

For DNT, an oncogen, the factor of 10 is inappropriate, so that

$$C_s = 2.35 \times 10^6 \ D_T$$

Timber-harvesting might involve perhaps 4% of the exposure to dust posed by other outdoor industrial activity; hence, the values of C_s would be higher by a factor of 25.

Ingestion of venison taken in hunting activity is assumed to be 11 kg by each member of a family of four per year. The value of K_{pa} is adjusted by a factor of 2/3, because venison is a leaner meat, perhaps 20% fat, than beef. The animals browse over a wide area, including uncontaminated land, for which reason a factor of 0.1 is introduced. The equation thus derived (with BW = 70) is

$$C_s = \frac{70 \times 365 \times D_T}{0.1 \times (2/3) \; K_{pa} \times 11} = 3.48 \times 10^4 \; D_T/K_{pa}$$

Results for the five scenarios examined in detail are shown in Table IV. It may be seen that subsistence farming and residential housing entail the most restrictive PPLVs at the Alabama AAP site.

Savanna Army Depot Activity (5)

The 5,330 hectare (13,170 acre) Savanna Army Depot Activity, north of Savanna, IL, consists of high ground and Mississippi River flood plain. In the flood plain are 223 hectares of waterways connected to the river; about 10 hectares of sediment plain in these waterways are considered potentially contaminated by munitions-related compounds (see Table I). Of these compounds, only TNT has been isolated (0.3 mg/kg in one sediment sample); DNT, TNB, and RDX are associated with TNT in other munitions contexts, hence they were also included. The waterways are fished by a number of activity personnel and retirees. These persons and their families may eat some of their catch, and thereby ingest those compounds that might be present in the fish (predominantly carp and catfish, both bottom-feeders). Acceptable safe sediment level guidance for these compounds was therefore desired.

The activity also has six bermed dry lagoons whose total area comprises 0.521 hectares. RDX has been found in surface soils of lagoons on high ground at levels up to 4,000 mg/kg. TNT and TNB have been found in groundwater beneath these high-ground lagoons at concentrations below 0.5 mg/L, and could be assumed to reside in lagoon soil, as could DNT. This groundwater is directed to the Mississippi River. There was concern that the leachate from the lagoons could pose a hazard in river-derived water supplies downstream.

The question of acceptable soil levels in waterway sediments was resolved by linking such levels to the human exposure route of fish ingestion. The fishermen involved do not require the fish they catch to provide a major portion of their diet. Thus, a safe-sided estimate of their fish dietary intake was set

Table IV. Allowable Concentrations (PPLVs) for Soil
Contaminants at Alabama Army Ammunition Plant

Scenario	Pathways Governing PPLV	PPLV for Soil (C_s) mg/kg)
Subsistence farming	Vegetable consumption	$953\ D_T$
Residential housing	Vegetable consumption	$953\ D_T$
Apartment housing	Soil ingestion by children	$1.2 \times 10^5\ D_T$
Industrial (outdoor worker)	Dust inhalation	$2.35 \times 10^7\ D_T$ ($2.35 \times 10^6\ D_T$ for DNT)
Hunting	Consumption of venison	$\dfrac{3.48 \times 10^4\ D_T}{K_{pa}}$
Timber harvesting	Dust inhalation	$5.9 \times 10^8\ D_T$ (5.9×10^7 for DNT)

at 0.01 kg per 70 kg person per day. This is about 1.5 times as high as the ingestion rate employed in Water Quality Criteria computations (13). Factors related to the pattern of intake by fish of the compounds of concern were introduced. These were: (1) The ratio of contaminated waterway bottom area/total waterway bottom area; (2) consumption by bottom feeders of detritus, i.e., 6% of their body weight (28); and (3) the depuration rate constants, k_2, of Table III. Based on these,

$$C_s = \frac{70 \times 223 \times k_2 \times D_T}{0.06 \times 10 \times 0.01} = 2.6 \times 10^6 \, k_2 \, D_T$$

Calculated values of C_s are shown in Table V. The C_s value for TNT (1.9×10^4 mg/kg) is well in excess of the concentration of TNT in the one measured sediment sample (0.3 mg/kg). It was expected that the C_s levels for other postulated compounds of concern would also be far in excess of sediment levels.

For leaching from lagoons, water consumption was considered the route of possible exposure to the pollutants. Estimated acceptable drinking water levels were determined by

$$C_w = \frac{BW \times D_T}{2 \text{ L/day/person}} = \frac{70 \, D_T}{2} = 35 \, D_T \text{ (mg/L)}$$

The concentration that could possibly be attained in the river due to contaminated dry lagoon soil (C_w) was then calculated. A worst-case scenario for delivery of pollutant to the Mississippi River would entail the assumptions that: (1) The rainfall on all lagoons becomes saturated with the contaminant; (2) all contaminated rainwater reaches the river, and (3) characteristic river flow is at an historic low.

Hence, the soil contamination levels do not enter into consideration, only the lagoon areas exposed to rainfall. Annual rainfall in the vicinity of the Activity is 0.86 m/year. The historic low flow of the Mississippi River in the Activity area was estimated at 2.5×10^{13} L/yr. Thus,

$$C_w' = \frac{\text{Lagoons area} \times \text{Rainfall} \times S}{\text{Historic low flow}}$$

Insertion of values for area, rain, and river flow, converted to a consistent unit basis, yields

$$C_w' = 1.79 \times 10^{-7} \times S \text{ (mg/L)}$$

The ratio C_w/C_w' may be considered a "safety factor;" if in excess of 1, it would indicate that the acceptable drinking

Table V. Calculated Values of Acceptable
Contaminant Levels in Waterway Sediment
(C_s) and Safety Factors (C_w/C_w')
for River-Derived Drinking Water,
Savanna Army Depot Activity.
(Based on Data of Table III;
for calculation, see text)

Contaminant	C_s (mg/kg)	C_w/C_w'
TNT	1.9×10^4	2.3×10^3
DNT	3.8×10^2	1.9×10^1
TNB	1.4×10^5	3.5×10^4
RDX	3.3×10^4	4.5×10^3

water concentrations would exceed the river contamination levels
for this worst-case scenario. This ratio, in terms of D_T and S,
is

$$C_w/C_w{'} = 1.95 \times 10^8 \times D_T/S$$

Values of $C_w/C_w{'}$ for possible compounds of concern appear in
Table V. All values are well in excess of 1, which indicates
that downstream drinking water supplies would not reach pollutant levels that might cause adverse human health effects.

Gratiot County Landfill (6)

Approximately 122,000 kg of polybromobiphenyls (PBBs) were
buried in the 40-acre Gratiot County, MI, landfill between 1971
and 1973. The upper natural clay barrier beneath the landfill
was breached in a few locations; hence, groundwater flowing
beneath the landfill can become contaminated as a result of
flooding of the landfill during periods when the groundwater
significantly rises, even if capping largely prevents leaching
by rainwater falling on the site. In addition to the landfill,
adjacent farms seem to have been contaminated by PBBs that may
have blown off trucks carrying such material to the landfill.
Although the solubility of PBBs, measured in distilled, deion-
ized or creek water, is below 10^{-3} mg/L, groundwater concentrations of up to 2.6×10^{-2} mg/L have been reported; this is not
surprising, since dissolved organic matter can greatly increase
the solubility of PBBs (25). Three land use scenarios have been
examined (see Table I); all rest on the assumption that the PBBs
will not be removed and that the landfill will be properly
capped. Other scenarios, in which PBB removal down to a safe
level was postulated, could be developed, and their consequences
explored.
   The transfer of PBBs from soil to plants is so low, e.g.,
Table III and References (6,29), that the only important issue
in the agricultural scenario appears to be soil ingestion (and
possibly ingestion of groundwater) by cattle. Based on an estimated half-life, $t_{1/2}$, in beef of 120 days (30) an estimated
mass of fat per animal, M_f, of 67 kg and a soil ingestion rate,
M_s, of 0.72 kg/day (31), a reasonably conservative soil-to-fat
bioconcentration factor can be obtained:

$$BF_s = \frac{0.72 \times t_{1/2}}{M_f \times 0.693} = 1.86$$

Where C_f = FDA guideline for PBB concentration in fat (26), the
SPPPLV for the soil ingestion pathway is then

$$C_s = C_f/BF_s = 0.3/1.86 = 0.16 \text{ mg/kg}$$

Note: Should groundwater (45.4 kg/day) be used for cattle, the
applicable PPLV would be

$$C_w = \frac{C_f \times M_f \times 0.693}{45.4 \times t_{1/2}} = 2.6 \times 10^{-3} \text{ mg/L}$$

If residences are supplied with well water, this would be
the most likely source of PBBs. For adults, the acceptable
concentration would be $C_w = 35 D_T = 1.3 \times 10^{-2}$ mg/L. (That for
children might be somewhat less.)
The residential soil concentration PPLV is governed by
children's soil ingestion, estimated at 10^{-4} kg/day (32).

$$C_s = \frac{BW_{child} \times D_T}{10^{-4}} = \frac{12 \text{ kg} \times D_T}{10^{-4} \text{ kg/day}} = 45 \text{ mg/kg}$$

If one source is assumed to be contaminated at less than
its applicable PPLV, then the PPLV for the other source need not
be reduced to zero. Thus, if the groundwater were contaminated
by 0.005 mg/L of PBBs, the residential soil PPLV, C_s, would be
(0.008/0.013) x 45 = 28 mg/kg.
The PPLVs applicable to industrial scenarios would possibly
be water ingestion (as in the case of residential housing), and
more likely dust inhalation. A conservative approach would be
to use the equation applied to DNT for worker exposure to dust,
i.e.,

$$C_s = 2.35 \times 10^6 D_T = 875 \text{ mg/kg}$$

In view of the above, groundwater in the vicinity of the landfill should be used as a drinking water supply only if the PBB
concentrations are vigorously monitored. Cattle grazing should
be restricted to the extent necessary.

Bangor Naval Submarine Base

Bangor Naval Submarine Base, on the Hood Canal in the State of
Washington, provides fine recreational facilities for service
people stationed there, as well as for civilian employees. A
proposal to divert runoff from munitions-contaminated areas
towards the recreational fishing pond, Cattail Lake, led to a
decision to identify hazard levels for the compounds of interest. In addition to trout, there was concern over contamination
of bivalves, such as oysters, cockles, and clams, at the pond's
outlet to Hood Canal. Bioconcentration factors (BCFs), assumed
applicable for both fish and bivalves, were developed for three
compounds (Table III). BCFs, together with D_T values and worst-
case levels of fish or bivalve consumption (0.4 kg/day) provided
PPLVs for the pond water, according to the equation

$$C_w = \frac{BW \times D_T}{\text{Consumption rate} \times \overline{BCF}} = 175 \ D_T/BCF$$

From this equation, C_w values were calculated, in mg/L, as:
TNT, 1.7×10^{-2}; RDX, 4.2×10^{-2}; PGDN, 1.6×10^{-2}. These very
stringent values reflect the lifetime consumption of almost a
pound of fish per person per day, and do not take into account
the fact that whole fish generally contain more fat than the
edible portions of fish or bivalves; the BCFs reflect whole fish
data. It is recommended that the foregoing C_w values be used as
detection limits for monitoring. If these are exceeded, the
assumptions may need to be reconsidered, since they appear to be
somewhat too stringent.

Summary

The examples provided above represent a variety of situations
where the uses to which land or water may be put would depend on
estimates of acceptable contaminant levels. Conversely, contaminants might be removed from land or prevented from reaching
water so that the land or water could be used beneficially for
specified purposes.

Literature Cited

1. Rosenblatt, D.H.; Dacre, J.C.; Cogley, D.R. "An
   Environmental Fate Model Leading to Preliminary Pollutant
   Limit Values for Human Health Effects," Technical Report
   8005, U.S. Army Medical Bioengineering Research and
   Development Laboratory, Fort Detrick, Frederick, MD, 1980,
   AD B049917L.
2. Dacre, J.C.; Rosenblatt, D.H.; Cogley, D.R. Environ. Sci.
   Technol. 1980, 14, 778-784.
3. Rosenblatt, D.H.; Dacre, J.C.; Cogley, D.R. "Environmental
   Risk Analysis for Chemicals"; Conway, R.A., Ed.; Van
   Nostrand Reinhold: New York, 1982, Chapter 15.
4. Rosenblatt, D.H.; Small, M.J. "Preliminary Pollutant Limit
   Values for Alabama Army Ammunition Plant," Technical Report
   8105, U.S. Army Medical Bioengineering Research and
   Development Laboratory, Fort Detrick, Frederick, MD, August
   1981, AD A104203.
5. Rosenblatt, D.H. "Environmental Risk Assessment for Four
   Munitions-related Contaminants at Savanna Army Depot
   Activity," Technical Report 8110, U.S. Army Medical
   Bioengineering Research and Development Laboratory, Fort
   Detrick, Frederick, MD, November 1981, AD A116650.

6. Rosenblatt, D.H.; Kainz, R.J. "Options and Recommendations for a Polybromobiphenyl Strategy in the Vicinity of the Gratiot County, Michigan Landfill," Technical Report 8204, U.S. Army Medical Bioengineering Research and Development Laboratory, Fort Detrick, Frederick, MD, 1982, AD A121243.
7. World Health Organization. "Evaluation of the Toxicity of a Number of Antimicrobials and Antioxidants," Sixth Report of the Joint FAO/WHO Expert Committee on Food Additives, WHO Tech. Rep. Ser. No. 228, pp. 9–11, 1962.
8. U.S. Environmental Protection Agency. "National Interim Primary Drinking Water Regulations," Fed. Regist. 40, pp. 59565–59588, December 24, 1975.
9. American Conference of Governmental Industrial Hygienists. "Documentation of the Threshold Limit Values," 4th ed., American Conference of Governmental Industrial Hygienists, Cincinnati, OH, 1980.
10. Cleland, J.G.; Kingsbury, G.L. "Multimedia Environmental Goals for Environmental Assessment," Vol. I. EPA 600/7-77-136a, Environmental Protection Agency, Washington, DC, pp. 60–61, November 1977.
11. Vettorazzi, G. in "The Evaluation of Toxicological Data for the Protection of Public Health;" Proc. Int. Colloq. Commission of the European Communities, Luxembourg, 1976, Hunter, W.J; Smeets, J.G.P.M., Eds.; Pergamon: Oxford, 1976; pp. 207–223.
12. Handy, R.; Schindler, A. "Estimation of Permissible Concentrations of Pollutants for Continuous Exposure," EPA 600/2-76-155. Environmental Protection Agency, Washington, DC, p. 61, June 1976.
13. U.S. Environmental Protection Agency. "Water Quality Criteria Documents; Availability," Fed. Regist. 45, pp. 79317–79378, November 28, 1980.
14. Lyman, W.J.; Reehl, W.F.; Rosenblatt, D.H. "Handbook of Chemical Property Estimation Methods;" Environmental behavior of organic compounds, McGraw-Hill, 1982.
15. Geyer, H.; Kraus, A.G.; Klein, W.; Richter, E.; Korte, F. Chemosphere 1980, 9, 277–91.
16. Kenaga, E.E. Environ. Sci. Technol. 1980, 14, 553–556.
17. Spacie, A.; Hamelink, J.L. "Recent Advances in the Prediction of Bioconcentration in Fish." Environ. Toxicol. Chem. 1982, 1, 309–20.
17. Spanggord, R.J.; Mill, T.; Chou, T.-W.; Mabey, W.R.; Smith, J.H.; Lee, S. "Environmental Fate Studies on Certain Munition Wastewater Constituents," Final Report, Phase I – Literature Review, Contract No. DAMD 17-78-C-8081, SRI International, Menlo Park, CA, 1980.
19. Hansch, C.; Leo, A.J. Substituent Constants for Correlation Analysis in Chemistry and Biology. John Wiley & Sons, New York, 1979.
20. Burlinson, N.E.; Personal Communication.

21. Berkowitz, J.B.; Goyer, M.M.; Harris, J.C.; Lyman, W.J.;
     Nelken, L.H.; Rosenblatt, D.H. "Literature Review –
     Problem Definition Studies on Selected Chemicals. Vol.
     III. Chemistry, Toxicology, and Environmental Effects of
     Selected Organic Pollutants," Final Report, Contract DAMD
     17-77-C-7037, Arthur D. Little, Inc., Cambridge, MA, June
     1978.
22. Banerjee, S.; Yalkowsky, S.; Valvani, S.C. Environ. Sci.
     Technol. 1980, 14, 1227-9.
23. Sullivan, J.H., Jr.; Putnam, H.D.; Keirn, M.A.; Pruitt,
     B.C., Jr.; Nichols, J.C.; McClave, J.T. "A Summary and
     Evaluation of Aquatic Environmental Data in Relation to
     Establishing Water Quality Criteria for Munitions-unique
     Compounds. Part 4: RDX and HMX," Final Report, Contract
     DAMD 17-77-C-7027., Water and Air Research, Inc.,
     Gainsville, FL, 1979.
24. Draft of "Recommended Interim Environmental Criteria for
     Six Munitions Compounds," prepared by Jack C. Dacre, circa
     August 1980.
25. Griffin, R.A.; Chou, S.F.J. "Attenuation of Polybrominated
     Biphenyls and Hexachlorobenzene by Earth Materials,"
     Environ. Geol. Note 87. Illinois Institute of Natural
     Resources, Urbana, IL, 1980.
26. Fries, G.F.; Cook, R.M.; Prewith, L.R. Dairy Sci. 1978,
     61, 420-425.
27. Clark, D.G.; Litchfield, M.H. Toxicol. Appl. Pharmacol.
     1969, 15, 175-184.
28. Leidy, G.R.; Jenkins, R.M. "The development of fishery
     compartments and population rate coefficients for use in
     reservoir ecosystem modeling. Appendix J. Digestive
     efficiencies and food consumption of fish." Final Report,
     Agreement No. WES-76-2, USDI Fish and Wildlife Service
     National Reservoir Research Program, Fayetteville, AR,
     1979.
29. Jacobs, L.W.; Chou, S.F.; Tiedje, J.M. Environ. Health
     Perspec. 1978, 23, 1-8.
30. Fries, G.F.; personal communication.
31. Healy, W.B. N.Z.J. Agric. Res. 1968, 11, 487.
32. Chaney, R.; personal communication.

RECEIVED April 15, 1983.

HUMAN RISK ASSESSMENT

Human Exposure and Health Risk Assessments Using Outputs of Environmental Fate Models

J. R. FIKSEL and K. M. SCOW

Arthur D. Little, Inc., Cambridge, MA 02140

The assessment of human exposure to and health risks associated with environmental pollution requires knowledge of ambient concentrations of pollutants in the air, water, and soil media. Based on the outputs of environmental fate and transport models, combined with available monitoring data, it is possible to estimate the average daily amount of a chemical ingested, inhaled, or absorbed by exposed human populations. This estimation may be performed at a national, regional, or local level, and may also identify specific subpopulations that are exposed to higher than average concentrations. Once the human exposure levels have been quantified, it is then possible to assess the risks of adverse health effects for various subpopulations. The per capita risk may be defined as the probability that an exposed individual will suffer a specified health effect either during or following exposure. Although both chronic and acute health effects may be addressed within this framework, attention will be focused upon methods for estimating carcinogenic risks based on extrapolations from laboratory animal dose-response data. Several successful applications of the above methodology will be presented, along with a discussion of the important assumptions, uncertainties, and limitations.

The purpose of an exposure and risk assessment is to characterize the magnitude and extent of human or environmental exposure to selected pollutants and to quantify the potential adverse effects of those exposures. The assessment can be used both to provide a baseline estimate of existing health risks attributable to an environmental pollutant and to determine the potential reduction in exposure and risk for various control options. Exposure and risk assessments are playing an increasingly central role in

0097–6156/83/0225–0287$06.25/0

providing Federal agencies such as the EPA with a quantifiable
basis for regulatory strategy development.   For example, the
agency's nationwide Ambient Water Quality Criteria for priority
pollutants and the Office of Drinking Water's SNARLs are based on
health risk; both TSCA and FIFRA require the conduct of some level
of exposure and/or risk assessment for certain cases.   This paper
will describe the methods currently available for exposure and
risk assessment, with particular emphasis on the use of fate model
outputs as a basis for human exposure and risk estimation.

    The scope of an exposure and risk assessment may be char-
acterized by a number of key features:

•   Geographic scale, which may be global, national, regional or
    local.

•   Pollutant sources, which may include industrial, residen-
    tial, commercial, both point and non-point sources; natural
    sources may also be included.

•   Environmental media, which may include air, surface water
    (water column and sediment), soil, groundwater, biota, or any
    combination thereof.

•   Pollutants addressed, which may be a specific compound or a
    class of related substances.

•   Receptor populations considered, which may include humans,
    animals, plants, micro-organisms, specific habitats or com-
    munities, or abiotic receptors; subpopulations that are
    exposed to unusually high pollutant levels may be high-
    lighted.

•   Exposure routes considered which may include ingestion,
    inhalation, dermal absorption, or any combination thereof.

•   Adverse effects considered, which may include acute or
    chronic human health effects as well as environmental ef-
    fects.

•   Time frame of the assessment, which may be retrospective,
    current or prospective.

•   Intended use of the assessment, which may be for regulatory,
    scientific, or public information purposes.

    An exposure and risk assessment will usually integrate a
number of different inputs, including health and environmental
effects evaluations as well as pollutant profiles for environmental releases, ambient monitoring data, and environmental fate

and distribution information. These inputs are combined with information about the number and distribution of exposed humans and other biota, and about the rate of intake for each exposure route (e.g., inhalation, ingestion, dermal absorption) through which these populations may be exposed to the pollutant in question.

The output of an exposure and risk assessment will usually describe the levels of exposure and quantity the population exposed for both humans and other biota, and will estimate the associated probabilities of the incidence of adverse health effects. Population exposure or risk, obtained by multiplying the individual (per capita) exposure or risk by the numbers exposed at each level of exposure, may also be a useful measure of impact. Various analyses can be performed on the results, for example, comparison of exposures in a particular geographic area against national average exposure levels. Likewise, for the same pollutant, environmental risks due to a particular industry might be compared against risks associated with occupational or household activities. In addition, the health risk of different substances could be compared for priority setting.

An important issue that must be recognized by practitioners is data adequacy and the associated levels of confidence in the exposure and risk assessment results. Depending on the accuracy and completeness of the required data, the results can range from well-defined numerical estimates to rough qualitative statements. Moreover, many of the techniques utilized to analyze data, notably fate modeling and dose-response extrapolation, involve a number of assumptions which may not be fully verifiable. Therefore, it is crucial that the outputs of the exposure and risk assessment are properly qualified in terms of model and data limitations. Despite such limitations, a well-organized and scientifically-documented assessment can be an extremely useful instrument for understanding pollutant impacts and guiding regulatory actions.

Exposure and Risk Assessment Methods. In evaluating exposure it is necessary to identify both the exposure route and the exposure medium; these are the two components of an exposure pathway. Exposure routes for humans include ingestion of food, water, soil, chemical products; inhalation of gases, water vapor and particulates; and dermal absorption, usually from solutions. Exposure media include finished drinking water, also foods such as fish, crops, plastic-wrapped or canned items; air in the vicinity of pollutant sources, in the workplace or home, in urban areas, and by highways; untreated surface water used for swimming and other forms of recreation, irrigation, or watering of livestock; and a variety of consumer products. Exposure pathways are logical combinations of these factors such as ingestion of surface water during swimming, or inhalation of air downwind from an atmospheric source.

Once the exposure pathways of concern have been defined, an
exposure and risk assessment can be performed.

In the following exposition, we will describe the exposure
and risk assessment methods in four major steps (See Figure 1):

- Pollutant Concentration Estimation: Use of Fate Models
- Exposure Route and Receptor Analysis
- Exposure Estimation
- Risk Estimation

An important input to the Risk Estimation step, as shown in
Figure 1, is the analysis of health effects associated with the
pollutant in question. Since environmental toxicology is itself a
complex and difficult field, we have confined this paper to a
discussion of how dose-response estimates can be utilized within a
risk assessment, with emphasis on human carcinogenesis. Thus, the
scope of this paper corresponds to the four steps surrounded by a
dashed line in Figure 1.

Pollutant Concentration Estimation: Use of Fate Models

A variety of modeling approaches may be used to estimate pollutant
concentrations in exposure media. These range from qualitative
estimates extrapolated from case examples or environmental scenarios, simple analytical equilibrium or transport models, to
complex multi-media models. In selecting an approach or approaches, it is important that:

- all exposure media of concern are included in the modeling
  analysis;

- the temporal resolution is equivalent for pollutant distribution and exposure estimation, and the time-steps are
  compatible with the dosing schedule of any effects studies
  (e.g., hourly SO_2 exposures);

- the spatial resolution of models takes into account the
  location and activities of receptor populations, and the
  results are at the same level of detail as the population
  breakdown;

- sensitivity analyses are performed for the environmental
  variables most significantly influencing exposure levels.

For many environmental situations, adequate models do not
exist or are just now under development. Furthermore, for new or
uncommon chemicals, many of the physical, chemical, and biological

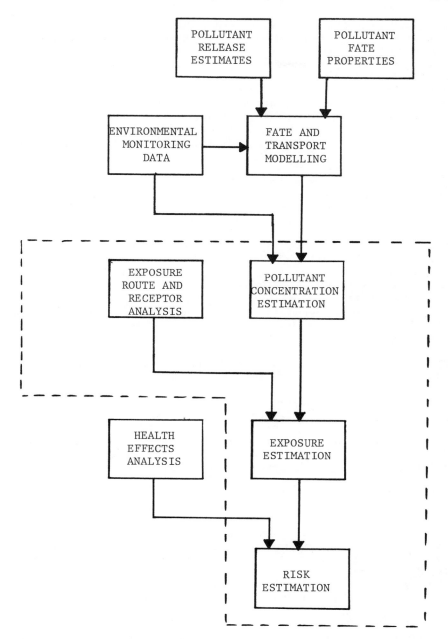

Figure 1.   Generic exposure and risk assessment methodology for environmental pollutants.

properties needed to estimate transformation rates, persistence, and distribution are not available. For example, few models exist to predict adequately the distribution of pollutants released from a landfill into ground water and surface water. Models to estimate residual concentrations of pollutants in edible foods, contaminated irrigation water, resulting from pesticide or nutrient application, or from dry deposition, are in very early stages of development. Therefore, uncertainties and limitations of the models should be identified, and estimates of pollutant distribution should be compared with monitoring data, if available.

Exposure Route and Receptor Analysis

The purpose of an Exposure Route and Receptor Analysis is to provide methods for estimating individual and population exposure. The results of this step combined with the output of the fate models serve as primary input to the exposure estimation step. Unlike the other analytic steps, the data prepared in this step are not necessarily pollutant-specific. The two discrete components of this analysis are: (1) selection of algorithms for estimating individual intake levels of pollutants for each exposure pathway; and (2) determination of the regional distribution of study area receptor populations and the temporal factors and behavioral patterns influencing this distribution.

Estimation of Intake.   For each exposure pathway, the general equation for estimation of the individual pollutant intake is:

(Pollutant Intake Rate) = (Exposure Medium Intake Rate)
                    • (Pollutant Concentration in Medium)

The exposure medium intake rate may be expressed simply as a unit mass per day (e.g., for ingestion), or with more detailed information, (e.g., for dermal absorption), as:

(Unit Exposure Medium Intake Rate) • (Frequency) • (Duration)
  • (Extent of Exposure) • (Pollutant Absorption Efficiency)

If possible, the intake should be expressed both as a statistical mean or median and maximum (e.g., 95th percentile). Ideally, a frequency distribution of exposure for the study area population is the goal. In most cases, however, the variability in exposure medium intake rates and pollutant concentrations are unknown and average/maximum values must suffice.

The time period to use in estimating and grouping pollutant intake values is determined by the eventual application of the exposure results. Exposure medium intake rates may vary on a daily (e.g., for inhalation) or seasonal (e.g., for recreational dermal absorption) basis. If the variability has a significant influence on pollutant intake rates, then exposure should be calculated in small enough time-steps to reflect these differences.

The absorption efficiency term allows estimation of the effective dose or the amount of pollutant which crosses the membrane of the exposed tissue (e.g., the lung) and reaches a target organ (e.g., the liver). For many pollutants this type of metabolic data is not available and consequently 100% absorption is a common preliminary assumption in exposure assessments. For well-studied substances such as radionuclides, a methodology for calculation of target organ doses has been developed for bone marrow, lungs, endosteal cells, stomach wall, lower intestine wall, thyroid, liver, kidney, testes and ovaries as well as for the total body.

Intake can be expressed either as a pollutant mass per unit time, as discussed above, or as a mass per kg of body weight per unit time. The latter expression facilitates comparison to health effects data, especially laboratory animal data, which are commonly reported in equivalent units. Similarly, depending on the route of exposure, intake may be estimated on an annual basis to address chronic effects, or on a smaller time scale for addressing acute effects including lethality, teratogenesis, reproductive and neurotoxic effects.

Table I illustrates the estimation of ambient exposures to benzene associated with various categories of atmospheric emission sources. Since benzene is a suspected carcinogen, annual exposure is an appropriate measure for assessing long-term effects.

The magnitude of exposure in a geographic area is a function not only of the amount of pollutant to which a "typical" individual is exposed but also of the size of the population exposed. This is especially important in the calculation of risk for an area or subpopulation. The resulting quantity is a population exposure factor which is the product of the individual pollutant intake level per unit time (average or maximum) multiplied by the population size exposed.

Population Characterization. An important part of any exposure assessment is the development of a detailed and up-to-date human demographic data base for the area being studied. These data can provide the basis for estimates of subpopulations associated with different exposure pathways. In national exposure assessments it is common to use an average population density for the total U.S. or to simply distinguish between rural and urban densitites. In a geographic exposure assessment in which site-specific data on pollutant releases, environmental fate and ambient levels are measured or estimated, it is important to have equally detailed population data. Population breakdowns by age, sex, housing and

Table I.   Estimated Ambient Exposure to Benzene

Source Category	Annual Exposure	
	10^6 Person-ppb-years	Percent
Chemical manufacturing	8.5	4.7
Coke ovens	0.2	0.1
Petroleum refineries	2.5	1.4
Auto emissions (SMSA's over 500,000 population)	150.0	82.5
Gas stations		
Nearby residents in urban areas	19.0	10.5
Self-service users	1.6	0.9
Total assessed	181.8	100.0

Source: U.S.E.P.A. (15)

other units are available from the 1980 U.S. Census; the data are grouped by state, county, tract, enumeration district/place, block group and block.

Within the general population, it may be important to quantify special subpopulations associated with specific types of exposure. The subpopulations can be divided into at least two categories: sensitive groups and specific exposure-related groups. The sensitive subpopulations are those that because of their age or health characteristics (e.g., very young or old age, poor health, high susceptibility to certain health effects) have the potential to be at greater risk when exposed. These groups can be identified within a geographic region and, therefore, associated with an exposure level for that area. Exposure-related groups may be broken down by drinking water supply, recreational activities (swimmers, fishermen, boaters), dietary habits (consumers of seafood, home garden produce), occupation, commuting patterns, consumer habits, specific subregions (near sources), and other important subdivisions.

Human Exposure Estimation

Exposure estimation is the next logical step in an exposure assessment. In this step, the data and methods developed in the previous steps are linked together so that the relationship between pollutant sources and human exposure can be examined. Through estimation of the degree of exposure rather than just estimation of concentrations in environmental media, a more detailed analysis of a pollution problem is possible, including:

• Estimation of the amount of pollutant to which a receptor is exposed in a unit time period, including a cumulative total for chronic exposure;

• Consideration of the influence of receptor behavior patterns or environmental conditions on receptor exposure.

For a limited number of exposure pathways (primarily inhalation of air in the vicinity of sources), pollutant fate and distribution models have been adapted to estimate population exposure. Examples of such models include the SAI and SRI methodologies developed for EPA's Office of Air Quality Planning and Standards (1,2), the NAAQS Exposure Model (3), and the GEMS approach developed for EPA's Office of Toxic Substances (4). In most cases, however, fate model output will serve as an independent input to an exposure estimate.

Ideally, an exposure assessment will represent the probable exposure of most of the local population for all times of the year and under all environmental conditions typical for the area.

Because an assessment of this detail can be very difficult and costly to perform, however, (especially if it requires the use of complex fate models), exposure estimates may be limited to average seasonal conditions (e.g., summer -- low flow -- low rainfall, etc.) and to specific "worst case" meteorological or hydrological conditions for that area (e.g., inversion conditions, 7 day-10 year low flow).

Human populations are likely to be exposed to a pollutant through more than one exposure route at a time. Total exposure may combine intake through ingestion of different substances, dermal absorption from surface water and water supply, and inhalation at different locations in the study area (e.g., work, home, recreational areas, commuting routes). Calculation of total exposure requires that the pharmacokinetics (absorption, metabolism, storage, excretion) for different exposure routes are understood for the pollutant of concern. Otherwise, only exposures by route can be combined.

Estimates of human exposure by route and subpopulation can be used directly, without comparison to health effects data, to evaluate potential pollutant problems in an area. For example, the following analyses can be performed:

•   Comparison of exposure levels to national "average" exposure levels;

•   Comparison of the impact of different routes or pathways of exposure such as drinking water vs. inhalation;

•   Comparison of locally attributable exposure vs. imports from outside (such as foods);

•   Comparison of day-time to night-time exposure or identification of seasonal variability in exposure;

•   Comparison of source-proximate (near-field) subpopulations to the rest of the population (far-field);

•   Identification of the sources responsible for the greatest exposure, both in terms of population affected and magnitude of individual intake;

•   Identification of sensitive subpopulations or high-exposure areas in the region being studied.

Particularly in the last two examples, fate models provide a useful tool not only for estimating concentrations, but also for tracing back the relative contributions of various sources to total exposure.

Risk Estimation

The final step in performing an exposure and risk assessment is to combine the information developed from analyses of exposure and effects in order to estimate the risk to humans. Risk may be defined as "the potential for negative consequences of an event or activity." In the present case, the event or activity is the release of a pollutant into the environment, and the negative consequences are adverse effects upon humans. Thus, if a pollutant is believed to be harmful and if it is present in the environment, there is certainly a potential for harm; that is, some risk exists. The purpose of a risk assessment is to go beyond such a qualitative statement, by estimating or measuring this potential. Although the nature of adverse effects is generally understood, the key difficulty in risk estimation lies in determining the <u>probability</u> that adverse effects will occur. The probability is comprised of two factors:

- the likelihood that groups of organisms will be exposed to potentially toxic concentrations of the pollutant;

- the likelihood that organisms will experience adverse effects given that they are exposed.

Different pollutants will present different types of problems within this framework, depending upon their properties and effects. For a highly persistent substance which is present in the human diet and known to have long-term effects, the main challenge lies in estimating the likelihood of adverse effects based on observed exposure levels. On the other hand, for a substance which degrades rapidly and appears only in scattered locations, but is known to be an acute toxicant, the focus should be on estimating the likelihood of exposure. Therefore, the risk estimation methodology must be flexible enough to encompass these and a multiplicity of other situations.

Given a population of susceptible individuals, risk may be expressed in several ways. One can state the probabilities that certain fractions of the population will be adversely affected (e.g., 5% chance that 9/10 will be affected, 20% chance that 1/3 will be affected). This is usually difficult to achieve. One can alternatively state the expected number that will be affected, allowing a certain margin for error (e.g., 1/3+25% will be affected). In the absence of such quantitative data, one can give an order-of-magnitude estimate which has no real measure of confidence attached to it (e.g., at most 1% will be affected).

Hence, in terms of level of precision, risk estimates may be classified into three types:

- Probability distribution
- Numerical interval
- Order of magnitude

As a rule, the level of precision of a risk estimate cannot exceed the precision of the exposure and effects data from which it is obtained. In the following we will focus upon carcinogenic risk estimation, for which it will often be possible to achieve at least interval estimates of risk.

Dose Response Thresholds. A dose-response curve can be defined as a relationship between the amount or rate of the chemical administered and the probability (i.e., risk) of the subject manifesting a delayed tumor at that dose. It is customary to estimate the "excess" risk; that is, the increase in probability of cancer above the normal background level. Hence the dose-response curve is a cumulative probability distribution function and should increase from zero to one, assuming that higher doses are more toxic. The estimation of carcinogenic risk using experimental data involves the formulation of a dose-response curve based on the observed effects. Toxicological data are generally expressed in terms of the percentage of organisms in which effects were observed at various dose levels. A number of alternate models are available which accept such data as input and calculate a dose-response relationship based on certain assumptions about the form of the curve. The purpose of this exposition is to discuss the use of data, the choice of models, and the handling of uncertainties in the process of carcinogenic risk extrapolation.

One important point of controversy in risk extrapolation is the existence of the threshold level for carcinogenic and muta-genic response to a pollutant. Some argue that an organism is able to cope with low doses of a substance through metabolic processes or repair mechanisms, so that harmful effects do not appear until a certain minimum threshold, or "safe dose", is surpassed. Others contend that a carcinogenic substance must be considered poten-tially harmful at any dose, and that even a single molecule may initiate a tumor at the cellular level. This is the so-called "one-hit" hypothesis.

The question of existence of a threshold has often been circumvented by the approach of selecting an "acceptable" risk level and determining the corresponding acceptable or "virtually safe dose" (VSD). However, as Upholt has pointed out, one need not assume a safe dose, nor insist on a zero tolerance (5). From a practical point of view, the behavior of the dose-response curve at low doses is an academic question, since there is an unavoidable background response due to the multitude of naturally occurring carcinogens, as well as the genetic heterogeneity of human populations. Moreover, due to realistic restrictions on sample size in animal experiments, it is extremely difficult to statistically reject the no-threshold hypothesis. Even with no observed tumors, the upper statistical confidence bound on the true risk will be linear at low doses, no matter what model is assumed (6). Thus the threshold issue may never be satisfactorily resolved through purely statistical arguments.

Dose-Response Data Conversion. The basis of any risk extrapolation for a specific carcinogen is a set of data, obtained experimentally or through field observation, describing the effects of that substance upon a population of organisms. Of course, human data are preferable for estimation of human risk, but epidemiological studies suffer from difficulties in quantifying exposures, as well as from a host of confounding factors, such as lifestyle, that may promote carcinogenesis (7). For these reasons extrapolations are more commonly performed using laboratory data for one or more species of mammals. Dose-response measurements are usually provided at several dose levels, including a control group which receives no dosage. The scientific quality and reliability of these data are an important consideration in risk assessment.

In assessing animal data, careful attention must be paid to the quality of the data, the incidence of spontaneous tumors in the control population, consistency if more than one study is available, and statistical validity. If the exposure route and experimental regimen employed do not agree with the most likely mode(s) of human exposure (e.g., intramuscular injection), the data must be interpreted cautiously. Consideration should be given to data on metabolism of the compound by the animal species tested, as compared with metabolism in humans if this information is known. If only in vitro data are available, only qualitative estimates may be possible because of uncertainties regarding the association between in vitro results and human or animal effects. The availability of associated pharmacokinetic data, however, may allow development of a rough quantitative estimate.

In order to extrapolate laboratory animal results to humans, an interspecies dose conversion must be performed. Animals such as rodents have different physical dimensions, rates of intake (ingestion or inhalation), and lifespans from humans, and therefore are expected to respond differently to a specified dose level of any chemical. Estimation of equivalent human doses is usually performed by scaling laboratory doses according to observable species differences. Unfortunately, detailed quantitative data on the comparative pharmacokinetics of animals and humans are nonexistent, so that scaling methods remain approximate. In carcinogenic risk extrapolation, it is commonly assumed that the rate of response for mammals is proportional to internal surface area

(7). Although other bases for conversion (e.g., body weight) have been utilized, the surface area method is the most widely accepted, and will be adopted here. This approach is more conservative, yielding risk estimates about an order of magnitude greater than those derived from scaling by body weight.

Assumptions in Risk Extrapolation. Risk extrapolation cannot be performed as a mechanical exercise, due to the need for judgment in the selection of data and application of dose-response models. In particular, there are a number of implicit assumptions inherent in risk extrapolation. They may be summarized as follows:

- It is assumed that a substance which is carcinogenic in laboratory animals is also a human carcinogen, although species differences in susceptibility are frequently observed.

- It is assumed the equivalent human doses can be scaled on the basis of relative surface area, although metabolic differences may be important in interspecies comparison.

- It is assumed that there is no absolute threshold for carcinogenic response, although detoxification mechanisms may prevent tumor initiation at low doses.

- If only ingestion experiments have been performed, it may be necessary to assume similar responses via the inhalation route, with appropriate dose scaling.

- It is assumed that the proportion of exposed humans who will experience tumors during their lifetime may be deduced from the proportion of laboratory animals exhibiting tumors at the time of sacrifice.

Given these assumptions, it is possible to apply various dose-response models that estimate response at low doses. Each model postulates a different shape of dose-response curve in extrapolating from high to low doses. By using several models, a range of uncertainty can be established between the least conservative and the most conservative results. We expect that the resulting range of uncertainty will dominate any of the uncertainties generated in the preceding models and, therefore, the uncertainty presented represents only the uncertainty generated in the risk estimates.

It is important to note that for a particular substance the assumption of carcinogenicity to humans may be false, even though it is a proven carcinogen in several animal species. In such a case, the lower bound on the excess risk to humans is effectively zero, in the sense that zero-risk is a possibility which cannot be dismissed.   Thus, the risk estimates obtained through dose-response extrapolation must be regarded as probable upper bounds on the true human risk.

Dose-Response Extrapolation Models.   A dose-response model is simply a hypothetical mathematical relationship between dose-rate and probability of response. For example, the simplest form of such a model asserts that probability of tumor initiation is a linear multiple of dose-rate (provided the dosage is well below the organism's acute effect threshold for the substance in question). In general, we will express dose-response models as follows:

$$P(x) = f(x; a,b,...)$$

where   x is the average daily dose (e.g., g/kg body weight/day)
             P(x) is the percent of organisms expected to respond at
             or below dose x,
             F( ) is a mathematical expression of the dose-response
             model
             a,b,... are fixed parameters of the model, dependent
             upon the experimental dose-response data.

Thus P(x) is a cumulative distribution function and will increase monotonically with dose-rate x.

There have been a number of recent survey articles and theoretical papers describing the available models for low-dose extrapolation. Through a literature review the most prominent models have been selected and discussed below. However, there are other models, less commonly used, that were not mentioned here for the sake of brevity. The models addressed below represent a good cross-section of the different features and capabilities that are pertinent to carcinogenic risk estimation.

The One-Hit and Multiple-Hit Models.   The one-hit model was one of the earliest dose-response models proposed for carcinogenesis, and has been recommended and utilized by several federal agencies for purposes of risk estimation for carcinogens (8). It is also widely used to quantify the carcinogenic effects of radiation (9). Though the scientific community has recently begun to favor the more general multi-stage model, the linear model is still preferred by many researchers due to its computational simplicity, its conservative nature, and the appeal of its underlying rationale.

The "one-hit" hypothesis states that the tumor initiation is a Poisson process, in that each additional molecule of a carcinogen produces an equal increment in the probability of a response, and that all such "hits" are independent. Consequently, the dose-response function is:

$$P(x) = 1 - \exp\left[-(ax + b)\right], \quad a > 0$$

where a is the "slope" of the low-dose curve and b is determined by the background response. At low doses, this expression is approximated by the linear function $P(x) = ax + b$.

This model does not allow for a threshold effect, and thus may overstate the risk at low doses. It tends to yield higher estimates of human risk than the other models discussed below.

The Log-Probit Model.   The log-probit model has been utilized widely in the risk assessment literature, although it has no physiological justification. It was first proposed by Mantel and Bryan, and has been found to provide a good fit with a considerable amount of empirical data (10). The model rests on the assumption that the susceptibility of a population or organisms to a carcinogen has a lognormal distribution with respect to dose, i.e., the logarithm of the dose will produce a positive response if normally distributed. The functional form of the model is:

$$P(x) = \Phi \left(a + b \log_{10} x\right)$$

where Φ is the normal cumulative distribution function.

This model tends to approach a zero probability rapidly at low doses (although it never reaches zero) and thus is compatible with the threshold hypothesis.   Mantel and Bryan, in applying the model, recommend setting the slope parameter b equal to 1, since this appears to yield conservative results for most substances. Nevertheless, the slope of the fitted curve is extremely steep compared to other extrapolation methods, and it will generally yield lower risk estimates than any of the polynomial models as the dose approaches zero.

The Multi-Stage Model.   The multi-stage model (not to be confused with the multi-hit model) is really a family of models in which the hazard rate is polynomial function of dose. First formulated by Armitage and Doll, it was later refined by Guess and Crump and their colleagues (11), (12). The underlying biological concept is that tumor initiation requires several successive stages or events at a particular receptor site. These stages are not simply "hits," but may involve biochemical activation processes coupled with cellular responses. The general functional form of the model with n stages is:

$$P(x) = 1 - \exp - \left[\sum_{i=0}^{n} a_i x^i\right], \quad a_i \geq 0$$

This model also reduces a one-hit model in the case n=1. However, when quadratic or higher-order polynomials are used, the shape of the curve changes considerably. Even so, at very low doses, provided that a_i 0, the linear component dominates. The resulting slope is usually much shallower than in the one-hit case, and thus yields a lower risk estimate.

Sample Application. The risk assessment techniques described above have been implemented for a number of suspected carcinogens, under a program sponsored by the Environmental Protection Agency's Office of Water Regulations and Standards. It was recommended that several models be implemented in a risk assessment application, since it is difficult to justify the use of any single model on scientific or statistical grounds. The various models represent a variety of unproven assumptions about the shape of the dose-response curve at low doses, and the resulting risk estimates may differ by several orders of magnitude. Thus the use of two or more models will provide an "envelope" of dose-response relationships, giving an indication of the wide uncertainty surrounding the true human carcinogenic response. The following criteria were used in the selection of the three models:

- Acceptance by the scientific and regulatory community.

- Ease and practicality of application with available data.

- Accommodation of competing hypotheses about threshold effects.

Due to the second criterion, time-to-tumor models were eliminated from consideration. These models require more detailed experimental data than is generally available. Moreover, it is difficult and unproductive to interpret the distribution of time-to-tumor in the context of human exposures. In most cases, the time-to-tumor variable would be integrated over a human lifetime, thus reducing the model to a purely dose-dependent one. Therefore we restrict our attention to quantal response models that estimate lifetime risks.

The most widely-accepted dose response model at the present time is the multi-stage model, which has great flexibility in curve-fitting, and also has a strong physiological justification. Although it is difficult to implement, there are already computer codes in existence that estimate the model parameters (13). The two most widely-used models, until recently, were the one-hit model and the log-probit model. They are both easy to implement, and represent opposite extremes in terms of shape – the former represents the linear non-threshold assumption, whereas the latter has a steep threshold-like curvature. In numerous applications with different substances it has been found that these three models span a broad range of risk estimates at typical human exposure levels. Their use in parallel will ensure a realistic portrayal of the great uncertainty associated with low-dose extrapolation.

The following example is based on a risk assessment of di(2-ethylhexyl) phthalate (DEHP) performed by Arthur D. Little. The experimental dose-response data upon which the extrapolation is based are presented in Table II. DEHP was shown to produce a statistically significant increase in hepatocellular carcinoma when added to the diet of laboratory mice (14). Equivalent human doses were calculated using the methods described earlier, and the response was then extrapolated downward using each of the three models selected. The results of this extrapolation are shown in Table III for a range of human exposure levels from ten micrograms to one hundred milligrams per day. The risk is expressed as the number of excess lifetime cancers expected per million exposed population.

Examination of these results reveals that the one-hit and multi-stage models are both linear in the low dose region, whereas the log-probit model increases steeply in a non-linear fashion through this region. The differences between the highest and lowest model predictions at each exposure level may be regarded as a measure of the model uncertainty in the risk extrapolation. For example, at 10 mg/day per capita exposure the range of model uncertainty is from 500 to 1000 cancers per million population, or a factor of about 2. At one hundred milligrams per day, however, the model uncertainty increases to a factor of about 4. This model uncertainty must be distinguished from the propagated uncertainty which might be present in the exposure estimates received from the exposure model. As with any risk assessment the results must be interpreted with extreme caution.

Limitations of Risk Assessment

There are a number of difficulties involved in the prediction of human health effects, which may be summarized as follows:

- scarcity of epidemiological or laboratory data concerning long-term effects of low level exposure to the hundreds of chemicals that may be hazardous;

- controversy about the biological effect mechanisms, resulting in a divergence of views about how to extrapolate the dose-response relationships obtained experimentally;

- high degree of uncertainty in the quantification of health effect incidence, due to both data and model error contributions.

Table II.   Carcinogenic Response in B6C3F1 Mice DEHP in the
             Diet for Two Years

	Dosage (mg/kg)	Equivalent Human Dose (mg/day)	Response*	Percent
Male Mice	0	0	9/50	18
	3000	1800	14/48	29
	6000	3600	19/50	38
Female Mice	0	0	0/50	0
	3000	1800	7/50	14
	6000	3600	17/50	34

*Hepatocellular carcinoma.

Source: National Toxicology Program, Carcinogenesis Bioassay of
           di(2-ethylhexyl) phthalate, (14).

Table III.   Probable Upper Bounds on Expected Excess Lifetime
              Cancers Per Million Population Due to DEHP Ingestion

	Exposure Level (mg/day)				
	.01	.1	1	10	100
One-Hit Model	1	10	100	1000	10,000
Log-Probit Model	–	.3	30	1000	20,000
Multi-Stage Model	.5	5	50	500	5,000

Source: Arthur D. Little, Inc.

The process of extrapolating high dose health effects to low-exposure risks for individual chemicals is confounded by many unresolved issues, including the difficulty of interspecies comparisons, and the uncertainty about the shape of the dose-response curve. An additional major source of uncertainty is introduced when one attempts to assess low-exposure risk in a multifactor situation. The etiology of the observed effect may defy interpretation. Often, it is not possible to determine the causative agent(s), and dose-response parameters are thus difficult to characterize. For example, exposure to two or more materials can enhance the cancer-inducing effects of each or perhaps only one component. Conversely, the actions of the toxicants may be antagonistic, either canceling each other out or perhaps slowing the onset of a response such that it may initially be overlooked. These synergistic or antagonistic interactions serve to complicate currently utilized risk extrapolation procedures, but they more accurately reflect real-world situations where the possibility for these interactions is large. As an example, certain polycyclic aromatic hydrocarbons (PAH) compounds are known to be promoters and/or initiators of carcinogenesis in rodents, so that a risk assessment must consider their presence in combination with other substances.

Some of the problems in extrapolating no-observed-effect levels in animals to the human situation are relatively easy to surmount by the use of scaling factors to compensate for differences in body weight or body surface area. In other cases, species differences may present difficulties in extrapolating between animal and human. For example, the structure of the rodent respiratory system is such that breathing through the nose is obligatory while in humans this is not the case. The result is a significant difference due to the protection of the rodent lung by the extremely efficient nasal filtering systems. Recognizing and compensating for anatomically determined differences between man and test animal requires biological sophistication and a cautious approach to extrapolation. The other major problem of interpretation of animal data concerns biochemical and pharmacokinetic diversity. If it can be demonstrated that a chemical is stored, absorbed, metabolized and excreted by the same pathways in animal or man, one can expect similar toxic consequences. If the pathways or conversion rates or excretion patterns are dissimilar, then predictions of the effect in man will be inaccurate. Once the effects analysis has been completed, perhaps only qualitative estimates of risk will be justifiable. Depending upon the compound, it may be possible to develop quantitative dose-response relationships and to extrapolate them to man as a function of exposure. Ideally, the analysis shoud identify exposure levels that are probably non-threatening, those levels that are associated with various effects, as well as the limitations of the data upon which these conclusions are based.

Literature Cited

1.   C. Anderson, C. Liu, H. Holman and J. Killus. "Human Exposure to Atmospheric Concentrations of Selected Chemicals." SAI Report EF-156R2, 1980.

2.   B.E. Suta. "BESTPOP: A Fine Grained Computer System for the Assessment of Residential Population." SRI International, 1978.

3.   T. Johnson and R. Paul, "NAAQS Exposure Model (NEM) and Its Application to Nitrogen Dioxide." Prepared for the U.S. Environmental Protection Agency OAQPS, 1981.

4.   L. Hall, "The OTS Graphical Exposure Modeling System (GEMS)." Pre-publication Draft, EPA Office of Toxic Substances, July 20, 1982.

5.   W.M. Upholt. "Models for Extrapolation of Health Risk" Proceedings of the EPA Conference on Environmental Modeling and Simulation. April 19-22, 1976, Cincinnati, Ohio.

6.   H. Guess, K. Crump and R. Peto. "Uncertainty Estimates for Low-Dose Rate Extrapolations of Animal Carcinogenicity Data." Cancer Research, 37, 1977, pp. 3475-3483.

7.   J.P. Leape. "Quantitative Risk Assessments in Regulation of Environmental Carcinogens." Harvard Environmental Law Review, 4, 1980, p. 86.

8.   Federal Register. Scientific Bases for Identification of Potential Carcinogens and Estimation of Risk: Request for Comments on Report, Federal Register, 44, 1979, pp. 39858-39879.

9.   National Research Council. The Effects of Populations of Exposure to Low Levels of Ionizing Radiation: 1980, National Academy Press, Washington, D.C.

10. N. Mantel and W.R. Bryan. "Safety Testing of Carcinogenic Agents." J. Natl. Cancer Inst., 27, 1961, pp. 455-470.

11. P. Armitage and R. Doll. "Stochastic Models for Carcinogenesis." Fourth Berkeley Symposium on Mathematical Statistics and Probability, University of California Press, Berkeley, CA, 1961.

12. H.A. Guess and K.S. Crump. "Low Dose Extrapolation of Data from Animal Carcinogenicity Experiments - Analysis of a New Statistical Technique." Math. Biosciences, 32, 1976, pp. 15-36.

13. K.S. Crump and W.W. Wilson. "A FORTRAN Program to Extrapolate Dichotomaus Animal Carcinogenicity Data to Low Dose." National Institute of Environmental Health Sciences. Contract No. 1-ES-2123. August, 1979.

14. National Toxicology Program. "Carcinogenesis Bioassay of di(2-ethylhexyl)-phthalate." (Draft Report) U.S. Dept. of Health and Human Services, DHAS Pub. No. 81-1773, 1980.

15. U.S. Environmental Protection Agency. Assessment of Health Effects of Benzene Germane to Low-level Exposure. Washington, D.C.: Office of Research and Development, 1978, EPA-600/1-78-061.

RECEIVED March 23, 1983.

INDEX

INDEX

Indexing by Susan Robinson
Production by Anne Riesberg
Jacket design by Kathleen Schaner

Elements typeset by The Sheridan Press, Hanover, PA
Printed and bound by Maple Press Co., York, PA

DATE DUE

FEB 7 1984 AP 25 98			
MR 25 85			
DE 18 87			
AP 18 '88			
MR 29 '90			
DEC 1 6 1989			
DE 14 '90			
AP 26 95			
MAY 1 1996			
DEC 0 2			
MAR 0 7 2011			